Steel Odyssey

In this wide-ranging interdisciplinary work, the authors draw on history, anthropology, and materials engineering to present a comprehensive and ambitious examination of the multifaceted roles of iron and steel throughout history and the current and future challenges faced by the steel industry.

Ohjoon Kwon, Joo Choi, and Hae-Geon Lee provide readers with an in-depth understanding of the history of iron and steel and their impact on human society from a materials engineering perspective. They begin by describing the characteristics of iron and steel and the history of human use of and interaction with these metals by compiling the fundamental knowledge necessary to understand iron's unique properties and metallurgical phenomena. Following this, they explain the influence of steel on human society and culture, focusing on the Industrial Revolution and warfare. They also give examples that are rarely discussed elsewhere, such as developments in Asia or iron's influence on thought and philosophy using Confucianism and Marxism as examples. Readers will then be able to apply this contextual knowledge to address the profound impact of emerging challenges, such as global environmental issues and the Fourth Industrial Revolution.

Despite the technical nature of this book, all terminology is fully explained to facilitate better comprehension for those who may not possess an engineering education or a direct interest in metallurgy. This book is therefore invaluable not only as a technical book but also as a guide to the development history of human civilization and its future challenges.

Ohjoon Kwon is a former Chairman and CEO of POSCO and is currently a representative of SEG. He obtained his PhD from The University of Pittsburgh in 1985. He has also held several chairmanships, including at the National Academy of Engineering of Korea, the Korea Iron & Steel Association, and the Korea–Australia Business Council.

Joo Choi is a former professor at the Graduate Institute of Ferrous Technology (GIFT), Pohang University of Science and Technology (POSTECH), Pohang, South Korea. He is a former Head of POSCO Research Laboratories. He received his PhD in 1995 from Seoul National University, Seoul, South Korea. He is a full member of the National Academy of Engineering of Korea.

Hae-Geon Lee is a professor emeritus of the Graduate Institute of Ferrous Technology (GIFT), Pohang University of Science and Technology (POSTECH) and an Honorary Professor at the University of Queensland, Brisbane, Australia. He received his PhD from the University of Washington, Seattle, WA, in 1983. He is a lifetime Member of Honor of the Iron and Steel Institute of Japan (ISIJ) in recognition of his significant contributions to the field of iron and steel.

Steel Odyssey

Tracing the Journey of Humanity Through the Lens of Steel

Ohjoon Kwon, Joo Choi, and Hae-Geon Lee

CRC Press is an imprint of the
Taylor & Francis Group, an **informa** business

Designed cover image: iStockphoto: lyash01

First edition published 2025
by CRC Press
2385 NW Executive Center Drive, Suite 320, Boca Raton FL 33431

and by CRC Press
4 Park Square, Milton Park, Abingdon, Oxon, OX14 4RN

CRC Press is an imprint of Taylor & Francis Group, LLC

© 2025 Ohjoon Kwon, Joo Choi, and Hae-Geon Lee

Reasonable efforts have been made to publish reliable data and information, but the author and publisher cannot assume responsibility for the validity of all materials or the consequences of their use. The authors and publishers have attempted to trace the copyright holders of all material reproduced in this publication and apologize to copyright holders if permission to publish in this form has not been obtained. If any copyright material has not been acknowledged please write and let us know so we may rectify in any future reprint.

Except as permitted under U.S. Copyright Law, no part of this book may be reprinted, reproduced, transmitted, or utilized in any form by any electronic, mechanical, or other means, now known or hereafter invented, including photocopying, microfilming, and recording, or in any information storage or retrieval system, without written permission from the publishers.

For permission to photocopy or use material electronically from this work, access www.copyright.com or contact the Copyright Clearance Center, Inc. (CCC), 222 Rosewood Drive, Danvers, MA 01923, 978-750-8400. For works that are not available on CCC please contact mpkbookspermissions@tandf.co.uk

Trademark notice: Product or corporate names may be trademarks or registered trademarks and are used only for identification and explanation without intent to infringe.

ISBN: 9781032727363 (hbk)
ISBN: 9781032727370 (pbk)
ISBN: 9781003419259 (ebk)

DOI: 10.1201/9781003419259

Typeset in Minion
by Deanta Global Publishing Services, Chennai, India

Contents

Nomenclature

AI	Artificial Intelligence
AHSS	Advanced High Strength Steel
AM	Arcelor Mittal
AMD	Acid Mine Drainage
BCC (bcc)	Body-Centered Cubic
BF	Blast Furnace
BOF	Basic Oxygen Furnace
BSC	Bethlehem Steel Corporation
CAL	Continuous Annealing Line
CC	Continuous Casting
CCS	Carbon Capture and Storage
CCUS	Carbon Capture, Utilization and Storage
CE	Carbon Equivalent
CEM	Compact Endless Casting and Rolling Mill
CGL	Continuous Galvanizing Line
CNT	Carbon Nano Tube
DP	Dual Phase
DRI	Direct Reduced Iron
EAF	Electric Arc Furnace
EDDQ	Extra Deep Drawing Quality
El	Elongation
ESB	Empire State Building
ESP	Endless Strip Production
FCC (fcc)	Face-Centered Cubic
FRP	Fiber-Reinforced Plastic
GA	Galvannealed
GI	Galvanized
GO	Grain Oriented
HCR	Hot Charged Rolling

HHA	High-Hardness Armor
HPF	Hot Press Forming
IC	Ingot Casting
IDA	Iron Deficiency Anemia
IEA	International Energy Agency
IF	Interstial Free Steel
LD	Linz and Donawitz
MOE	Molten Oxide Electrolysis
N(G)O	Non-(Grain)Oriented
NSC	Nippon Steel Corporation
OWTC	One World Trade Center
PVD	Physical Vapor Deposition
RHA	Rolled Homogeneous Armor
SC	Strip Casting
SFE	Stacking Fault Energy
SMR	Small Modular Reactor
TKS	Thyssen Krupp Stahl
TRIP	Transformation Induced Plasticity
TRL	Technology Readiness Level
TS	Tensile Strength
TSC	Thin Slab Casting
TWIP	Twinning-Induced Plasticity
UAV	Unmanned Aerial Vehicle
U-AHSS	Ultra Advanced High Strength Steel
USS	United States Steel Corporation
WHO	World Health Organization
WSA	World Steel Association
WSD	World Steel Dynamics
WTC	World Trade Center
X-AHSS	eXtra Advanced High Strength Steel
YS	Yield Strength

Foreword

STEEL CAN BOAST OF an extensive literature to which many have contributed, but every so often, a gem emerges with a brilliance that consumes the reader. It is in this context that I pay homage to the authors of *Steel Odyssey*, which true to its title, is an epic written by giants of the subject. But these giants are benevolent in that they have nurtured both the knowledge and practice of steels.

This book does not cover just the mere thousands of years that humans have recognized iron in one of its forms. It harks back to the origins of the universe. The process in which hydrogen evolves into heavier elements terminates at iron, because energy generation by the fusion process stops there. Even heavier elements are created in massive explosions that occur during the imbalance between gravity and thermal energy. These heavy elements over time decompose by fission toward iron. So in principle, our galaxy will eventually turn into just iron. This book begins with this story, explaining why the Earth has in excess of 10^{23} kg of iron at its core, with snippets of how humanity uses and abuses iron.

Iron has to be treated as royalty, but purely out of reverence. With iron, the atomic structure, the crystal structure, indeed the structure at a variety of length scales can make or break the purposeful objects that are integral to our everyday life. And its beautiful metallic bond ensures that atoms in its solid form can slide over each other without the material losing integrity. This makes iron and its alloys safe engineering materials, capable of absorbing and storing energy before fracture. This book introduces the concepts in a way that the reader can communicate them with friends and family. For example, the variety of magnetic phenomena associated with both electrically conducting liquid iron, and its different solid states not only protects life on Earth but permits ingenious artificial devices that can drive motion.

But making iron is not easy. The first oxygen generated on Earth was consumed by dissolved iron in the oceans, turning it into oxide. Only then did O_2 reach the atmosphere to sustain creatures. The iron must therefore be recovered from the oxide. The technology for doing this evolved rather slowly until Bessemer, who sparked the first industrial revolution. The processes then went in leaps and bounds to today, where new ideas are being explored with vigour. Who knows what the next decades will bring to the reduction of iron oxide.

No one would argue about the role of iron in civilization, but its earliest contributions were pernicious in supporting warfare. This continues to be the case, but the vast majority of iron produced today is used for the benefit of all in the world. Just look around you and imagine a world without iron. This book ventures deep into the historical significance of iron, from the ancient Anatolia to the Aryans and modern art such as the Cloud that adorns Chicago.

The industrial revolution sparked the creation of powerful machines, from the steam engine to the "iron horse," the metal ships and Elon Musk's starship for anticipated journeys to Mars. Some failed disastrously, such as the Liberty ships in which the steel was not compatible with the effects of welding. A nicely illustrated section of this book shows what can be done reliably and safely.

I do not want to delve too much into steel and war, for there are few justifiable wars. But I admit, it is inevitable that war features in the use of steel, from chains, cannons, the creation of empires, and so on. Almost anything invented by science and technology will make it into some form of weapon, be it a material or software. All of science has a "dual use" in modern jargon. But space travel, for example, would be impossible without steel and look at how it fires the imagination. All this is dealt with in this book.

This is followed by the modern history of steel in different geographical locations, a story worth reading to understand the barriers to building steel plants, and the passions that overcame those hurdles. In India, the creation of Tata Steel was opposed by the Colonial Governor of India, but Jamshedji Tata proved more tenacious. In South Korea, Tae Joon Park could not at first able to summon help to create the steel industry, but he persisted. So with aid from Japan, we now have POSCO which all would recognize as one of the creative steel industries. Indeed, the final chapter describes the unique FINEX process which may be adaptable to reduce powdered oxide using the lightest element on Earth.

I would like to finish by saying a few words to honour the authors. Ohjoon Kwon has an enviable record of leading one of the most inventive steel companies in the world, while at the same time providing long-term support to academia working on steels. Joo Choi, after heading the research laboratories at POSCO, is now a professor at the heart of all things ferrous. Hae-Geon Lee has an admirable record in academia, with selfless devotion to caring for education in metallurgy. This praise cannot ever be enough, but trust me, having these authors write this book gives us an adventure that none should not miss.

Harry Bhadeshia
Professor of Metallurgy, Queen Mary University of London
Emeritus Tata Steel Professor of Metallurgy, University of Cambridge

Foreword: At the Crossroads of Personal and Civilizational History

It is a paradox that the former POSCO Chairman Kwon Ohjoon, a soft-spoken man with a delicate touch, has spent his life with iron, a rough and tough metal. In fact, iron is an alchemist's favorite material. Its metamorphosis and transformation are endless. That's why iron is central to the history of civilization, the history of warfare, the history of industry, and the history of thought. Readers may call this an exaggeration, but millions of years of human history across the globe bear it out. Iron is commonly perceived as harsh and frightening, but when one observes its nature and characteristics, one quickly realizes that it has been a gentle, intimate, and indispensable companion of human life.

The appearance of POSCO, one of the main contributors to Korea's industrialization, is overwhelming and majestic. The production process of the steel plant, which operates around the clock, is a river connected by thousands of kilometers of pipes and rails. Elements and knowledge fuse together to create a spark, and steel cascades down from the end of the spark. The image of an old-fashioned forge is so strong that steelworks are often perceived as noisy, cluttered workplaces, but they are, strictly speaking, "imagination power plants." The 20th-century French philosopher Gaston Bachelard identified the four elements of human imagination as water, fire, earth, and air, and iron is a divine creation that fuses all four.

This means that human imagination and the creative space of iron are homologous.

This book is an interdisciplinary experience of the steelworker written from that perspective. As the top-level experienced experts of metallurgy, the authors' knowledge and understanding of iron is, dare I say, century old. The authors' analytical and contemplative abilities to soften the resolute majesty of steel with a historical and humanistic perspective gained in the workplace. Their capability to bring steel dissolved in intellectual observation into the space of historical and philosophical understanding and to touch the essence of human civilization is comparable to the innovative wonders of the historic Bessemer process.

For the authors, iron is not just iron. It is a gift from the gods that has sculpted and carved centuries of civilization and made the foundations of human life solid. Readers are encouraged to set aside the fact that the authors of this book include the CEO of one of the world's leading steel companies. To do so would be to miss the point of this book. This book is the travelogue of a steel scientist exploring the inner workings of steel, the exploration of a steel archaeologist discovering its enduring connection to human civilization, and the testimony of countless moving scenes witnessed in the production process and on the shop floor. Not a single shot is left untouched and reborn into a new scene by the author's detailed interpretation.

What is the meaning of a lifetime spent with iron? This question is the gate that guides the life history of an individual to the history of human civilization. After leaving the work site, the authors boldly exit and walk the path of civilization created by iron. The encounter between iron and man thousands of years ago was a remarkable meeting. The civilization scholar J. Diamond wrote a history of continental civilization in response to a very simple question. In his international bestseller, how did whites live a life of superiority over indigenous peoples? The world's best-selling book "Guns, Germs, and Steel" was born in this way. Iron is a common element of guns, germs, and steel. Guns and steel are made of iron, and iron is also an important element of fungi. Iron is also the absolute element that governs the origin of life. DNA, the world's most mysterious life form, is a complex chemical reaction of carbon, hydrogen, nitrogen, and carbon, with sulfur, phosphorus, calcium, and iron involved in the process. Iron also permeates the origin of life.

The lead author Dr. Kwon's gaze wanders from the relationship between life on Earth and iron to the creation of the universe. Iron is at the origin of the mystery that breathed life into him before he was born. The mountain behind his childhood home was named Mt Chultan meaning the mountain containing iron, and his nickname was Chulwu meaning the iron friend, so iron was his lifelong companion and destiny. Iron is nature itself and the lever that combines nature and labor to create the richness of human life. The author's journey begins with a personal history and moves with the civilizational history of iron.

It is not surprising that iron and ironware were behind the creation of a civilization where surplus production occurred, the family system was established, and the unfamiliar concept of private property was created, but the authors' raison d'être acquires a new meaning when it is linked to their lifelong profession. Iron gave birth to capitalism, and it also gave birth to socialism as a critical alternative to it. It was its art and weapon. This ambivalence, captured by senior managers and the top scientists who have spent their lives working with iron, is laid out like a trap in the text, captivating the reader.

There is no need to add the lead author's achievements as a CEO. He was the first in the world to commercialize the high Mn steel and the FINEX process at POSCO, and he created a new industrial engine based on the importance of lithium. What about smart factories? Readers may ask. A steel company is smart? The answer is yes. I have witnessed it myself. Smartization, a word that is probably not registered in the Oxford dictionary, is a reality at POSCO's steel plants. In 2019, POSCO was selected as a Lighthouse factory by the World Economic Forum (WEF). Lighthouse factory, isn't it cool? We invite readers to go into the story space between.

It is inspiring that top executives of a steel company and professors in the university have worked together to write such a heartfelt book that starts with scientific historical observations on the history of steel civilization and leads to the touching story of the Korean steel industry. The authors are neither the civilization scholars nor the historians. The authors' encouragement for millennials at the end of this book is heartwarming. This book is not an autobiography, but I can't quite fathom why the broad perspective of civilization, autobiographical experience, scientific inquiry, and generational concerns resonate so strongly. I look forward to reading this book at the intersection of personal and civilizational history.

Ho Keun Song

Endowed Chair Professor, Hallym University, South Korea

Former Endowed Chair Professor, Postech, South Korea

Former Endowed Chair Professor, Seoul National University, South Korea

Preface

IRON PERMEATES OUR LIVING spaces to such an extent that its omnipresence is evident at first glance. Its ubiquity is a profound testament to the vital role it plays in our lives, enabling a level of convenience and efficiency that would otherwise be unimaginable. Throughout the course of civilization, iron has steadily risen to an irreplaceable position, fundamentally shaping the production of indispensable necessities, and profoundly influencing human activities. As we find ourselves immersed in the information and communication age of the 21st century, iron's exceptional electromagnetic properties continue to drive its applications to new heights.

In this book, we explore the ever-evolving nature of iron and its profound impact on the tapestry of human history and civilization. With meticulous scrutiny, we analyze and interpret a wealth of evidence to unravel the intricate story of iron's evolution. Through our meticulous research, we shed light on the transformative influence it has exerted throughout the ages and delve into the depths of its significance. We seek to uncover the dynamic relationship between iron and the evolution of human society and illuminate the interconnectedness that has shaped our collective journey

This book begins with an introductory section (Chapter 1) that provides the reader with an overview of steel and its beneficial characteristics for human beings such as strength, plastic workability, magnetic properties, and weldability. It then delves into the origins and formation of iron in the universe and introduces how the Iron Age began with a comprehensive compilation of the fundamental knowledge necessary to manufacture iron from ore (Chapter 2).

Advances in steelmaking technology are described in detail in Chapter 3. Over the course of more than 3,000 years, humans have devoted themselves to developing technologies and applications for iron. However, significant advances in the mass production of steel have occurred primarily in the last 150 years. Before the advent of converter refining in Europe in the mid-19th century, mass production of steel was impractical and costly. Subsequently, the invention of oxygen blowing to the converter in the mid-20th century revolutionized steel production, making it much more affordable and accessible.

The impact of iron on human civilization and society is undeniably significant, with numerous tangible examples to support this notion. These examples are briefly examined in Chapters 4 and 5. Iron's influence on thought and philosophy is more subtle and indirect, making it difficult to detect at first glance. However, if iron has indeed influenced human civilization and culture, it is reasonable to assume that it has also shaped human thoughts and, consequently, social norms. To shed light on the less obvious effects of iron, an attempt has been made to infer its influence on thought and philosophy by focusing on these indirect consequences, using Confucianism and Marxism as examples. While the steam engine is often credited as the catalyst for the Industrial Revolution, it is crucial to recognize that the production of steam engines themselves depended heavily on the availability of steel. By exploring these indirect effects and linking them to key philosophical and ideological frameworks, we can begin to understand the far-reaching influence of iron and steel on human thought and society

Undoubtedly, one of iron's most prominent impacts on human history has been its association with warfare, a topic discussed in detail in Chapter 6. War, driven by the innate human desire to triumph in competition, has played a central role throughout history. Wars are perpetually waged because the victor has the power to shape the course of human civilization. It is no exaggeration to say that from ancient times to the present day, wars have been won or lost based on the possession of abundant iron resources and the ability to forge superior weapons from them. Thus, nations seeking triumph in warfare have prioritized the advancement of steel-making technologies and the production of high-performance steel weapons. The history of humankind is inextricably intertwined with the history of war, and at the heart of that history lies the profound influence of steel. The constant quest for military superiority has been intertwined with advances in iron and steel technologies, driving nations to seek innovation and sophistication in their armaments. Thus, the historical narrative of humankind is intricately woven with the tapestry of warfare, and at its core lies the unbending influence of steel. The development of steel-based weaponry and the quest for military supremacy have left an indelible mark on the course of human history. Acknowledging this central connection allows us to grasp the profound impact of steel in shaping the destinies of nations and the evolution of our collective story.

Chapters 7 and 8 discuss the present and future of the steel industry. Discussing the future of the steel industry, in particular, presented significant challenges. The sheer number of variables involved created a sense of pressure when attempting to forecast future developments. To make accurate predictions, it was imperative to gain an accurate understanding of the current situation, identify existing challenges, and explore imaginative alternatives for solving them. Since global warming is a serious issue in the current situation, it is attempted to replace the existing blast furnace operation with a new hydrogen reduction process in the global steel community. Moreover, the Fourth Industrial Revolution which includes advances such as artificial intelligence and big data on human history cannot be overlooked. Recognizing this, the potential consequences and opportunities of emerging technologies have been carefully examined.

This book has been written by three authors: Dr. Ohjoon Kwon, Professor Joo Choi, and Professor Hae-Geon Lee. Over the course of nearly half a century, the three authors have cultivated a deep knowledge in the field of iron and steel at POSCO and its affiliated university, Postech. With unwavering dedication and a strong passion for technological advancement, management innovation, and talent development, they have effectively fulfilled their responsibilities. After accumulating many years of invaluable experience and making significant contributions to the steel industry, all three have recently retired from active service.

Acknowledgments

DR. OHJOON KWON, THE lead author and a former CEO of POSCO, is grateful to POSCO and its employees, including the former CEOs for their consistent encouragement in recognizing the purpose of this book. He is immensely grateful to the co-authors, Dr. Joo Choi and Professor Hae-Geon Lee, for their participation and dedication. Dr. Kwon would also like to express his appreciation to Mr. Bong-Rak Son, CEO of TCC, and Mr. Young-Chul Hong, CEO of KISWIRE, for their kind encouragement and support.

All three authors would like to express their sincere gratitude to their wives and other family members for their unwavering encouragement. Numerous individuals provided substantial assistance throughout the manuscript writing process, including past and present employees with whom Dr. Kwon had the privilege of working at the POSCO Group, as well as many professors actively involved in steel research at universities. Although it is difficult to express individual gratitude in this context, he would like to take this opportunity to express his sincere appreciation to each of them.

CHAPTER 1

Characteristics of Steel

1.1 THE INTERACTION BETWEEN IRON AND HUMANITY

Iron is the most abundant element on the Earth where we live on.[1] It is the fourth most abundant element in the Earth's crust, and as we go deeper into the Earth, its amount increases, making up most of the core. Human civilization has developed along with an understanding of steel and iron, which is the nascent stage of steel. "Steel," the subject of this book, is the best innovative material developed by mankind over a period of more than 3,000 years through artificial processes such as adding carbon or alloying elements to iron and heat treatment. Steel is a strong, hard, and tough material. Since the Industrial Revolution, it has been the most commonly used material in many of the structures that we encounter in our daily lives, such as buildings, bridges, automobiles, machinery, and home appliances. Iron is also essential in our bodies, where it carries oxygen through the blood, and in plants, where it plays an important role in the production of chlorophyll which releases oxygen.[2]

Iron constitutes only 0.004% of our bodies,[3] but it plays a crucial role in sustaining our lives. Hemoglobin, a component of our blood, binds with oxygen inhaled into the lungs and carries it to various organs in our body, generating the energy necessary for sustaining life. The reason why our blood appears red is due to the presence of iron in our red blood cells. Iron deficiency can cause symptoms of anemia, and in severe cases, can even be life-threatening. Iron deficiency anemia (IDA) is by far the most common form of anemia worldwide. The World Health Organization (WHO) estimates that nearly 2 billion people or 25% of the world's population are anemic, and about half of them suffer from IDA.[4]

DOI: 10.1201/9781003419259-1

Iron located at the center of the Earth creates a magnetic field around the Earth that protects living organisms on this planet from harmful radiation emitted by the Sun.[5] This magnetic field also enables wireless internet communications, allowing people to enjoy the benefits of the information age today.[6] Steel is used in almost every structure we encounter in our daily lives, ranging from giant bridges, airports, and opera houses with large spaces to super-tall buildings and enormous ocean structures. (See Figure 1.1). Steel not only makes automobiles, trains, and railways but also serves as the core material for ships, expanding the range of human life and mobility, and plays a significant role in the production and transportation of various energy sources such as gases and oils. Iron is also used as a key component material in various audio and video devices, making various cultural activities possible.

FIGURE 1.1 Various applications of steel products. (a) Grain storage complex, with permission from Posco International, (b) Street lights, courtesy of https://www.publicdomainpictures.net/en/free-download.php?image=empty-road-with-streetlights&id=366369, (c) Vehicles, (d) Washing machines, (e) Kitchen utensils, courtesy of https://unsplash.com/ko/%EC%82%%EC%A7%84/%EC%8A%A4%ED%85%8C%EC%9D%B8%EB%A0%88%EC%8A%A4-%EC%8A%A4%ED%8B%B8-%EC%88%9F%EA%B0%80%EB%9D%BD%EA%B3%BC-%EC%88%9F%EA%B0%80%EB%9D%BD-BYm1kkFasEI, (f) High-pressure vessel, (g) Elevator, (h) Electromagnet, courtesy of https://www.needpix.com/photo/283497/, (i) Steel cans, courtesy of https://www.freepik.com/free-vector/preserve-tin-realistic-set_6128950.htm#query=steel%20cans&position=31&from_view=keyword&trk=ais, (j) Manhole cover, (k) Free cutting steel, (l) Motor core, with permission from Posco International. (Continued)

FIGURE 1.1 CONTINUED Various applications of steel products. (m) Ship, courtesy of https://unsplash.com/ko/%EC%82%%EC%A7%84/rot-weisses-frchtschiff-neben-dcourtesy of k-iFvGi2Jcbrc, (n) Wind power tower, with permission from Posco International, (o) Airplane engine, courtesy of https://commons.wikimedia.org/wiki/File:Daimler_DB_605_cutaway.jpg, (p) Locomotive and railroad, courtesy of https://ko.m.wikipedia.org/wiki/%ED%8C%8C%EC%9D%BC:Mugunghwa-ho.jpg, (q) Suspension bridge, courtesy of https://commons.wikimedia.org/wiki/File:George_Washington_Bridge_from_New_Jersey-edit.jpg, (r) Steel structures for construction, (s) Bridge and structure, courtesy of https://unsplash.com/ko/%EC%82%%EC%A7%84/ponte-de-metal cinza-4SDCaKV-bR0, and (t) Offshore structure, with permission from Posco International.

Recently, the use of steel in electric cars, which is rapidly gaining popularity, has increased significantly. This is due to the increased use of Fe-based electric components of various sizes and types used in motors and sensors, along with ultra high strength steel (UHSS) to reduce vehicle weight. Stainless steel home appliances and cookware also an important role in preserving fresh food and healthy cooking, thereby helping to prolong human life.

The discovery of iron by ancient humans is attributed to meteorites. A meteor, known in astronomy as a meteoroid, is a space rock that enters the Earth's atmosphere and becomes extremely hot due to air resistance and oxidation. The meteor we see in the sky is not a stationary star, but a shooting star, moving at speeds of up to 10–18 km per second. Most meteoroids are destroyed when they enter the Earth's atmosphere, but some can withstand the heat and successfully land on Earth. These types of meteoroids are called meteorites.

Meteors are divided into three types depending on their composition: stony, iron, and stony–iron. Meteorites that contain a large amount of iron are

FIGURE 1.2 '"Hoba," Fe-16% Ni, the largest iron meteorite on Earth discovered in 1920.[8] (Courtesy of Wikimedia Commons, https://commons.wikimedia.org/wiki/File:Hoba_meteorite_(15682150765).jpg)

called iron meteorites or siderites. Iron meteorites also contain 4–20% nickel and 0.3–1.6% cobalt.[7] Figure 1.2 shows the largest iron meteorite found on Earth, the "Hoba" meteorite, which fell in Africa about 80,000 years ago.[8]

The oldest known iron artifact is a set of meteorite beads found in a tomb in Egypt that dates back to 5,200 years ago (Figure 1.3).[9] The meteorite dagger (Figure 1.4) found in the tomb of King Tutankhamun, who ruled in the 1360s BC, is also a famous iron artifact.[10] The blade of the sword had hardly rusted, which is a characteristic of iron meteorites that contain a large amount of nickel. A dagger made of meteorite dating from 2400 to 2300 BC, even earlier than this dagger, was excavated at Anatolia Alcahöyük Grave K.[11, 12]

Humans first encountered iron through meteorites, which sparked their fascination with it. Meteorites have a composition similar to the elemental composition of the Earth's interior, with iron and nickel as their main components. The iron carried by meteorites was a very rare material that was delivered from the sky, making it very expensive. In the market of the ancient Assyrian people 4,500 years ago, iron was about eight times

FIGURE 1.3 Photograph of three of the excavated nine iron beads from Gerzeh, Lower Egypt. From left UC10738, UC10739 and UC10740. (With permission from Petrie Museum of Egyptian and Sudanese Archaeology, UCL, ©Rob Eagle journal.)[9]

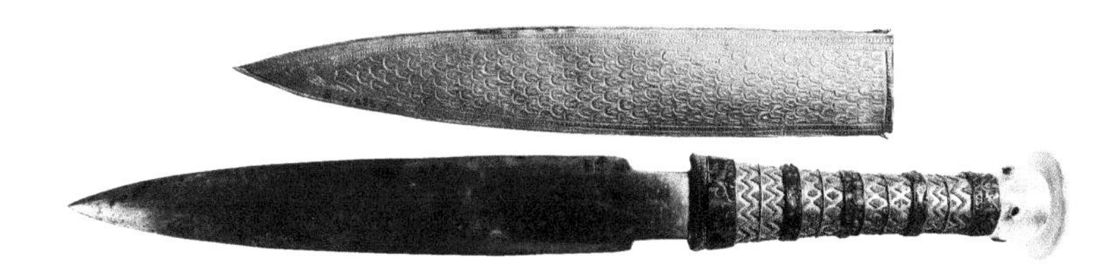

FIGURE 1.4 Golden scabbard and meteorite dagger found in the "Pyramid of Giza."[10] (With permission from Copy Clearance Center.)

more expensive than gold. At that time, ancient people regarded iron as a "gift from the gods" and called it by the name that symbolized the sky and fire.[13] The ancient Egyptians referred to meteorites as *bia-en-peb*, the Sumerians called them *an-bar*, the Hebrews called them *parzil*, and the Hittites called them *ku-an*. All of these names indicate that people were aware that iron was a gift from the universe or a star in the sky that fell to Earth.[13]

The first humans to produce iron from iron ore were in Kaman-Kalehöyük, modern-day Anatolia, around 2000 BC.[14] This ironmaking technology was passed down to the Hittite Empire, which produced iron from iron sand obtained from a nearby river in Hattusa in Anatolia around 2000 BC. Strengthened by ironmaking and steelmaking technology, the

Hittite Empire rapidly expanded its territory and competed even with Egypt for regional dominance in the eastern Mediterranean. After the collapse of the Hittite Empire, the use of iron and steel in agriculture and warfare spread to Europe and Africa between 1000 BC and 500 BC and was transmitted to China, Korea, and Japan through India and Central Asia. The use of iron and steel ushered in the Iron Age, which brought about significant changes in culture, politics, and society. Iron was also involved in the rise and fall of ancient Greece and the Roman Empire and had an impact on the expansion of the Islamic world.[15–17]

The Old Testament of the Bible mentions iron more than 40 times in History Books, including Genesis. It is recorded that iron was used to make chariots, which were the equivalent of tanks, and had a significant impact during wars. In the Book of Judges, King Jabin had 900 iron chariots, which he used to threaten Israel with an army of about 10,000 soldiers, but General Barak, who received the grace of Yahweh, destroyed Jabin's army. In addition to weapons, iron is mentioned several times in the Old Testament as a tool for agriculture and daily life.[18]

Humanity has been smelting iron for a long time, mainly through the bloomery and blast furnace operation. From ancient blacksmiths to 17th-century ironworks, charcoal was the main fuel used to make iron.[15–17] In England, a situation arose where massive deforestation occurred due to the harvesting of wood, the raw material for charcoal. To address this issue, a method of smelting iron using coke, a carbonized form of coal, was developed and improved.[19, 20] This coke-fueled ironmaking using coke led to the mass production of pig iron and greatly contributed to the Industrial Revolution. Subsequently, the Bessemer converter and open-hearth steelmaking technologies, which could mass-produce steel with superior characteristics were developed one after another. The converter technology was established through the Thomas converter and then the LD converter, which used pure oxygen, allowing for the simultaneous improvement of steel productivity and quality.[21]

Britain's iron and steel production accounted for half of the world's production in the mid-19th century, making it the leading steel-producing country and playing a pivotal role in driving the Industrial Revolution. The position of the British Empire, which had led numerous colonies, was transferred to the United States in the late 19th century.[22, 23] From the beginning of the 20th century until after World War II, the United States dominated the world with its excellent steelmaking capabilities, not only in the construction of skyscrapers and bridges but also in automobile production and weapons manufacturing.[24] However, history has come full circle, and the roles of the US and Europe in the steel industry have been taken over by East Asian countries such as Japan, Korea, and China. Japan was very strong in terms of technology,[25–27] Korea in cost competitiveness,[28] and China in mass steel production.[29] In developing countries with rapid population growth and urbanization, such as certain regions in India, South Asia, and Africa, steel consumption is increasing faster than at the current pace, and global steel consumption is expected to continue to increase until at least 2050.[30] On the other hand, the emission of carbon dioxide and pollutants due to the mass production of steel and the environmental degradation from large-scale mines are being seriously discussed, and measures to improve them are actively adopted.

The steel industry's evolution has been a catalyst for ongoing industrialization and the advancement of information technology. Steel is still widely used, in fact, its usage is increasing. Such enduring significance has led some to propose that the era of the Iron Age is still going on today. In the future, steel will play an even greater role in meeting the new demands that we continue to create. Today, 90% of the metals produced are steel. From this perspective, "Tracing the Journey of Humanity Through the Lens of Steel" will be an intriguing task, which requires knowledge not only of the origins, characteristics, and technological evolution of iron and steel but also of the impact of steel on human civilization and society.

1.2 TYPES AND APPLICATION OF STEEL

Iron is typically classified into pure iron, steel, and pig iron based on its carbon content. Both steel and pig iron contain a relatively high amount of carbon, with pig iron being classified when the content is higher than 2% and steel when the content is lower. Pure iron is characterized by almost no impurities such as Si, Mn, P, or S, and can be produced through the electrolysis process. Pure iron has good malleability and ductility at room temperature, but its low tensile strength precludes its use as a structural material. However, it is widely employed as an electrical material for transformers and generators due to its high magnetic permeability characteristics.[6] Physical properties of pure iron include a specific gravity of 7.87 g/cm^3, a melting point of 1,538°C, thermal conductivity of 80.4 W/m•K, and Brinell hardness of 20 to 30 BH.[31]

The pig iron produced in the blast furnace (BF) contains a large amount of carbon, typically between 3.5 and 4.5%. Pig iron is used as the raw material for castings. It has little malleability and ductility, which makes

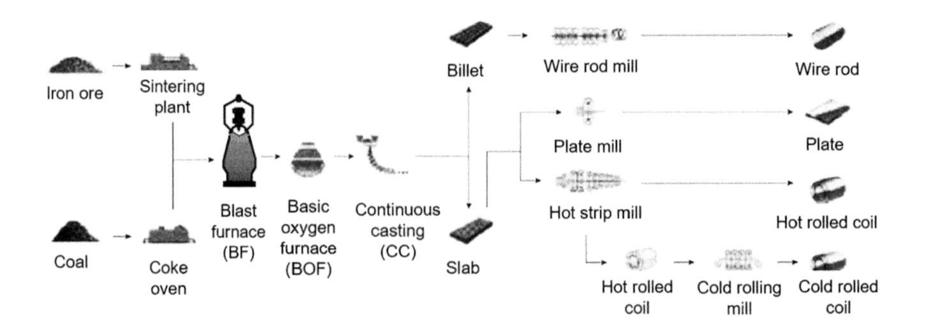

FIGURE 1.5 The steel manufacturing processes in the integrated steelworks.

it difficult to shape. In most of the existing integrated steelworks, as seen in Figure 1.5, sintered iron ore is reduced by coke in a BF to make molten iron. Coke serves as a reducing agent as well as a heat source. It also serves as a carburizer during the iron ore smelting process, resulting in pig iron with a carbon content of about 4%.

Molten iron from the blast furnace, also known as hot metal, is converted and refined into steel by removing carbon and impurities. In the ordinary integrated steel mills refining is carried out in a basic oxygen furnace (BOF), as well as through a secondary refining process such as a ladle furnace and vacuum degassing. Steel is produced by decarburizing the pig iron in the BOF until the carbon content is lowered to 0.04% or less, and then adding carbon to a desired level. The molten steel is then cast into semi-finished products such as slabs, blooms, and billets through continuous casting (CC). The typical steel manufacturing processes in modern integrated steel mills are depicted in Figure 1.5.

The semi-finished products are reheated and processed by rolling to produce the finished steel products shown in Figure 1.6. Steel products are classified by appearance into hot- and cold-rolled coils, sheets, plates, wires, and bars according to their appearance, as shown in Figure 1.6. Sheets are produced by cutting hot-rolled or cold-rolled coils. Hot-rolled steel coils are rolled thinner to make cold-rolled steel coils at room temperature. Because cold-rolled steel after continuous annealing has good formability, it is widely used in automobiles and home appliances with complex shapes. Wire rods and bars are produced by continuously rolling blooms or billets, and they have a wide range of applications.

Coated steel sheet is a product in which other metals such as zinc and tin are thinly coated on the surface of the steel sheet to prevent corrosion.

FIGURE 1.6 Steel products of various shapes. (a) Hot and cold rolled coils, With permission from World Steel Association, https://worldsteel.org/images/world-steelposcosouthkoreahotrolling12jpg-272/, (b) Sheet, With permission from World Steel Association, https://worldsteel.org/images/worldsteelposcosouth excavated koreacoldrolling8jpg-303/, (c) Plates, With permission from World Steel Association, https://worldsteel.org/images/worldsteelposcosouthkoreahot rolling31jpg-314/, (d) Round bars, Courtesy of https://www.freepik.com/free-photo/large-steel-factory-warehouse_1243108.htm#page=2&query=steel%20ro excavated und%20bar&position=2&from_view=search&track=ais, (e) H beams, With permission from from Hyundai Steel, https://www.hyundai-steel.com/kr/products-technology/products/wideflangebeams.hds?mobile=pc, (f) Wire, With permission from World Steel Association, https://worldsteel.org/images/world excavatedsteelposcosouthkoreahotrolling24jpg-292/.

Zinc-coated galvanized steel is mainly used for automobiles, electronic products, and interior and exterior construction materials. Tin-plated steel is primarily used for can-making. The plate is wider and thicker than the steel sheet. It is produced in a rolling mill dedicated to manufacturing plates, and its thickness is usually more than 6 mm. The heavy plate is widely used as an industrial material for large bridges, high-rise buildings, oil storage tanks, and oil tankers. Long steel products such as wires and bars are produced by reducing the thickness of blooms & billets by rolling at high temperatures. The long steel products are often named according to the shape of the cut section, such as H bar, angle, A channel, and T bar. They are mainly used for framing or foundation work such as bridges, buildings, factories, and subways. Wires and rods go through a subsequent rolling process to make very thin piano wire or tire cord, and various sizes of bolts, nuts, springs, needles, and nails. For example, in the case of tire cord wires used in automobile tire manufacturing, it is important to minimize the diameter of the wires to reduce fuel consumption, while maximizing strength to improve lifespan, stability, and ride comfort.

Steel is classified into carbon steel and alloy steel. Carbon steel elongates well when its carbon content is low and is mostly widely used for structural applications. The typical carbon content ranges from 0.002 to 1.0%. Carbon is a very inexpensive and effective element for strengthening. The addition of carbon increases the hardness of the steel, making it stronger and tougher when properly heat-treated. The carbon content of tire cords often exceeds 0.8% where the tensile strength is greater than 4,000 MPa. For automotive applications, very low carbon IF (interstitial-free) steels are widely used where the carbon content is extremely low (about 0.002%) and the elongation (El) is greater than 50%. Therefore, the control of the carbon content is a crucial factor in steel production at a reasonable cost.

Carbon steels are the most widely used steel products, especially for structural applications such as buildings, vehicles, bridges machinery, and household goods. In the case of steel-framed buildings, we do not notice the steel because most of the steel is buried in the concrete form of a frame or reinforcement. Although more than half of the weight of the automobile body we drive or ride in every day is made of steel, the steel is not immediately visible on the surface, because it is hidden in the paint or non-ferrous materials. We may ignore the ubiquitous existence of steel even though it is so widely used. It is recommended to recall the final product of steel in the steel plant to better recognize the crucial role that steel plays in our lives.

Alloy steel contains special metal elements such as nickel, chromium, manganese, tungsten, or molybdenum and is a value-added product. Alloy steel is divided into low-alloy steel and high-alloy steel. Low-alloy steel typically has a total alloy content of less than 2% and is widely used for structural purposes. High-alloy steels, in which the total amount of alloy added sometimes exceeds 30%, are widely used for aerospace, military, and chemical industries. They are also used for buildings, kitchen utensils, and household appliances that require special functions, including stainless steel with high corrosion resistance.

High-alloy steel, which contains large amounts of alloys such as nickel, molybdenum, and chromium in large amounts, is used for special purposes such as heat, corrosion, and wear resistance. The steel is used in power generation, mining, aviation, pressure vessels, and nuclear power. A typical example is stainless steel which contains about 10% or more chromium and has excellent corrosion resistance. A thin passivation film is formed on the surface of steel, which inhibits the further oxidation of steel by making it difficult for ions to move between the iron and the corrosive environment. Due to this property, stainless steel has a long-lasting appearance and is widely used as a kitchen utensil loved by cooks.

The lattice structure of stainless steel changes depending on its composition and is typically classified into three types: austenitic, ferritic, and martensitic. There is also duplex stainless steel in which austenite and ferrite exist together. Austenitic stainless steel, which accounts for over 70% of products, has good workability and weldability. Ferritic stainless steel is less expensive due to little or no addition of nickel and is highly resistant to corrosion at high temperatures, making it in high demand for vehicle exhaust systems and gas purification containers. Martensitic stainless steel, which has a high carbon content, is suitable for applications requiring high strength and high hardness.(32)

High-alloy steels for aerospace and armor are high value-added products. These include maraging steel, which has excellent mechanical properties at both room and high temperatures and is used for rocket outers, engines, and airplane landing gears.(33) Armor steel requires hardness and is often manufactured by single or multi-layer rolling of martensitic alloys.(34) Other special steel alloys include shape memory alloy steel(35) made by mixing titanium with nickel, a sponge alloy(36) made to float on water with a foaming agent, a damping alloy(37) that does not generate sound even when vibrated, and a heat-resistant alloy(38, 39) that can withstand high temperatures up to 1,000°C.

In addition to developing high value products, reducing manufacturing costs and improving quality are crucial in the steel industry. With the advent of big data and artificial intelligence (AI), efforts are actively underway to make the steel industry smart with digital transformation, a highly desirable trend expected to significantly contribute to cost reduction and quality improvement. This is part of the so-called Fourth Industrial Revolution of the steel industry.(40)

1.3 IRON PHASE TRANSFORMATION

Compared to other materials, steel is particularly superior in four characteristics: such as strength, plastic workability, weldability, and

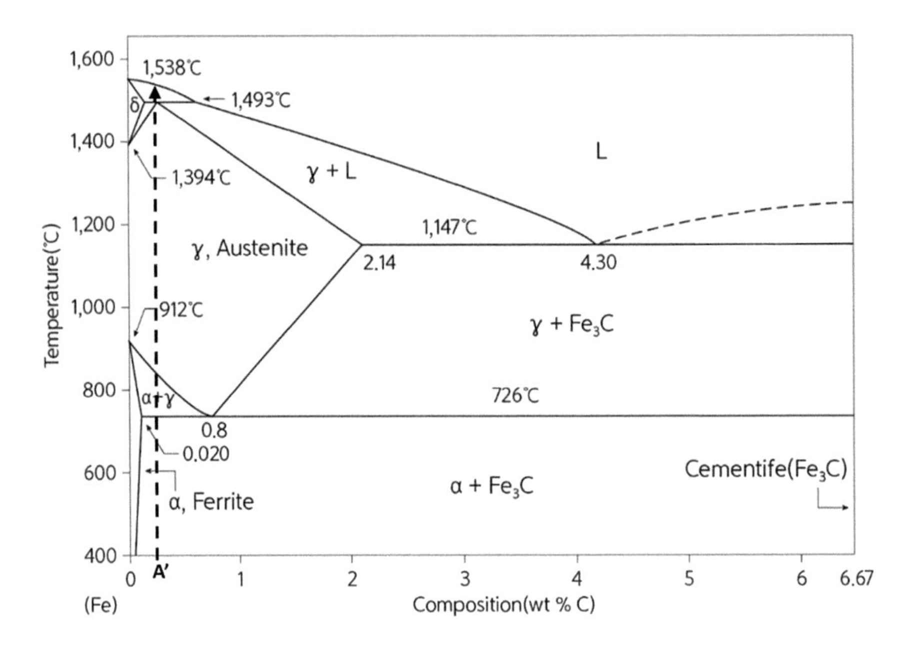

FIGURE 1.7 Fe–C phase diagram.

magnetism. These characteristics are closely related to the microstructure, and the microstructure changes variably depending on the processing conditions such as alloying, heating, cooling, and plastic working. These phenomena can be understood through the Fe–C phase diagram shown in Figure 1.7.[41]

Changes in crystal structure according to the carbon content and temperature lead to changes in material properties. In the case of pure iron, which is located at the far left of the phase diagram, the crystal structure at room temperature is the α-ferrite phase of body-centered cubic (BCC or bcc) (Figure 1.8). When the temperature rises to 912°C or higher, as shown in Figure 1.7, the austenite γ-phase of face-centered cubic (FCC or fcc) becomes stable. The arrangement of Fe atoms in BCC and FCC is schematically shown in Figure 1.8. When the temperature rises further to 1,394°C or higher, δ-ferrite which is BCC again becomes the stable phase.[41–44]

In the case of 0.2% carbon steel A', which is widely used as structural steel, α-ferrite and carbide called cementite (Fe_3C) appear as the stable and metastable phases, respectively at room temperature. When the temperature is raised to 727°C or higher as shown by the arrow in Figure 1.7,

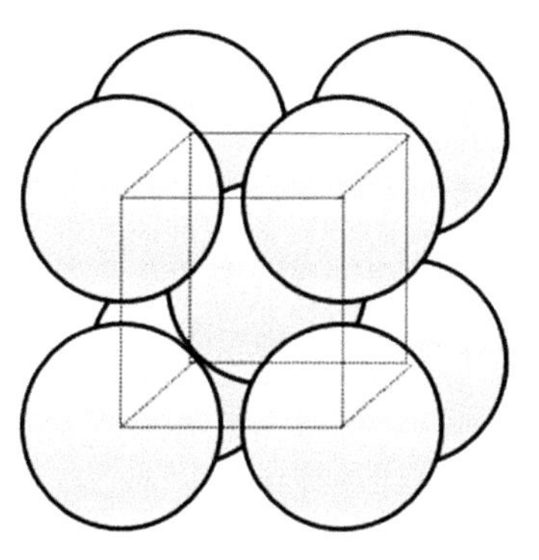

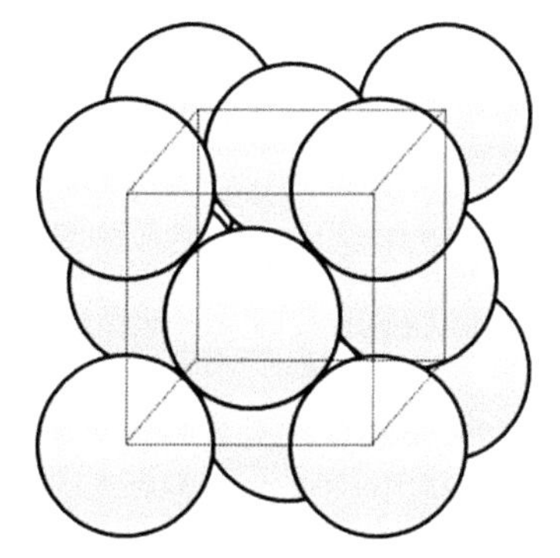

FIGURE 1.8 Crystal structure of iron.

a mixed phase of α-ferrite and austenite appears, and when heated higher, a structure with 100% austenite is formed. When the temperature is raised again, the phase transformation proceeds, and the stable phase changes from $\gamma \Rightarrow \gamma + \delta$-ferrite $\Rightarrow \delta$-ferrite + liquid $\Rightarrow$ 100% liquid. Since the properties of steel depend entirely on the microstructure, the desired microstructure must be obtained by appropriately controlling heat treatment and alloying.

Figure 1.9 shows that austenite transforms to ferrite at 912°C when the temperature is lowered. This phase transformation leads to an increase in volume or a decrease in density.(45) It is interesting to note that this feature of austenite transformation on cooling is unique to steel, and is contradictory to what is expected from the nature of the volume-temperature

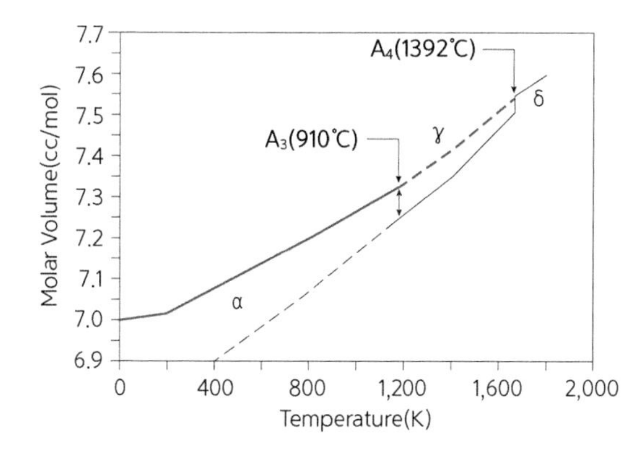

FIGURE 1.9 Phase transformation and volume change on heating pure iron.

relationship. It is quite natural that the volume contracts and the density increases on cooling. This phenomenon is a critical factor in determining the degree of strengthening when steel is transformed, which will be discussed later. It is also noted that the denser FCC has high toughness with its closed packing structure, and it can be used as a material for cryogenic gas containers. Steels using this principle are stainless and high Mn austenitic steels, which will be discussed in Chapter 8.[46]

1.4 BENEFICIAL CHARACTERISTICS OF STEEL

A brief explanation of the terms used is given here. Tensile Strength (TS) is the strength at which the maximum load appears in the strength-elongation curve when it is pulled with a tensile tester (see Figure 1.10). In the initial stage of pulling, the test specimen returns to its original size like a spring. However, when the strength exceeds a certain limit, the specimen does not return to its original size. This limiting strength is called Yield Strength (YS). The degree to which a material stretches until it breaks is called elongation (El). The specimen begins necking when the load reaches a maximum and eventually breaks. The elongation at the maximum load is called the uniform elongation (El_U) and the elongation at break is called the total elongation (El_T).

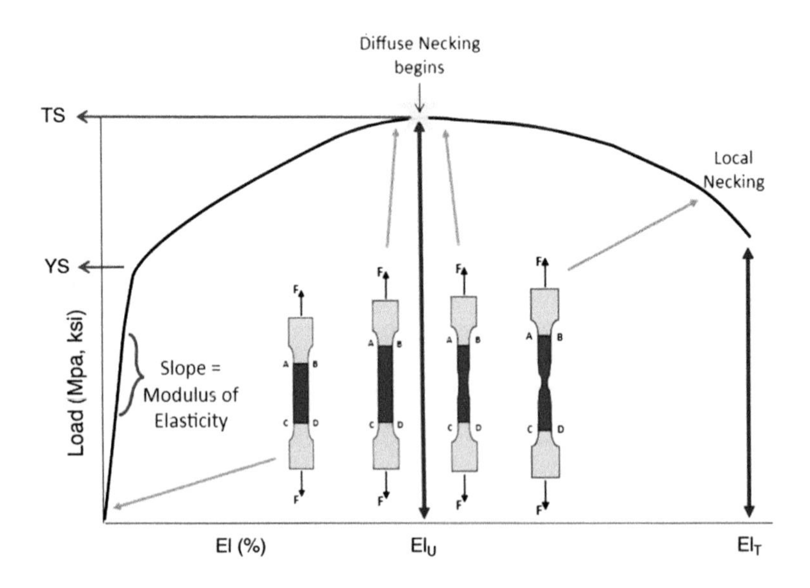

FIGURE 1.10 Strength–elongation curve from which mechanical properties are derived.

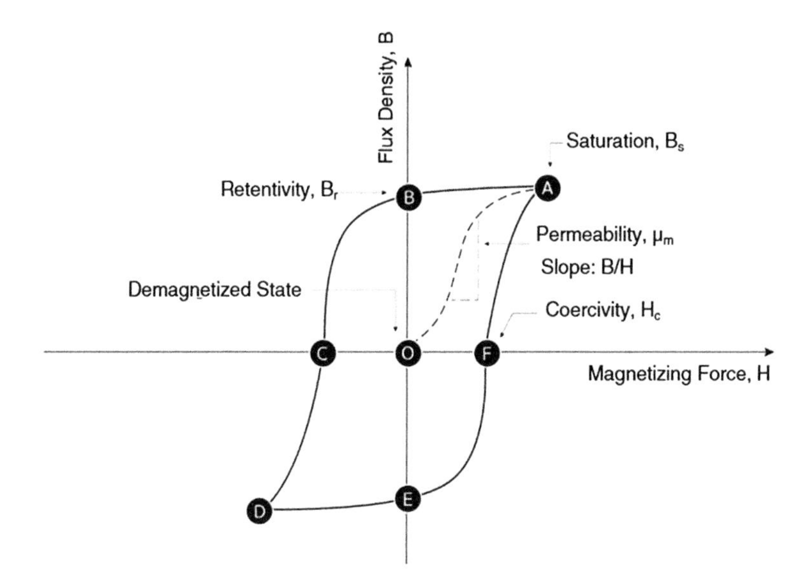

FIGURE 1.11 Magnetic hysteresis loop including B–H curve and permeability.

Hardness is defined as the resistance to indentation. Hardness is measured in various ways, and the Brinell hardness is determined by measuring the result of how much the material is indented when a ball of a certain diameter is impressed.

Permeability (μ) is the ability of a material to become magnetized when exposed to a magnetic field (see Figure 1.11). If B is the flux density of a magnetic body in a space with a magnetizing force of H, then the permeability, μ is defined as B/H.

1.4.1 Strength

There are various metals on the Earth, but steel is the most widely used among them because it offers excellent cost efficiency along with good mechanical properties. Figure 1.12 shows the TS ranges of various ferrous products compared to those of two non-ferrous metals, aluminum, and titanium. The TS of steel shows a much wider range than aluminum and titanium, ranging from 200 MPa to 5,000 MPa. Why is the range of steel's TS wider than that of other metals? To answer this question, we need to look at the dislocation movement and the strengthening mechanisms of metals.

The regularity in the crystalline metal often breaks and generates defects in the crystal during plastic deformation. When a linear irregularity occurs in a crystal the defect is called the dislocation. It is well accepted

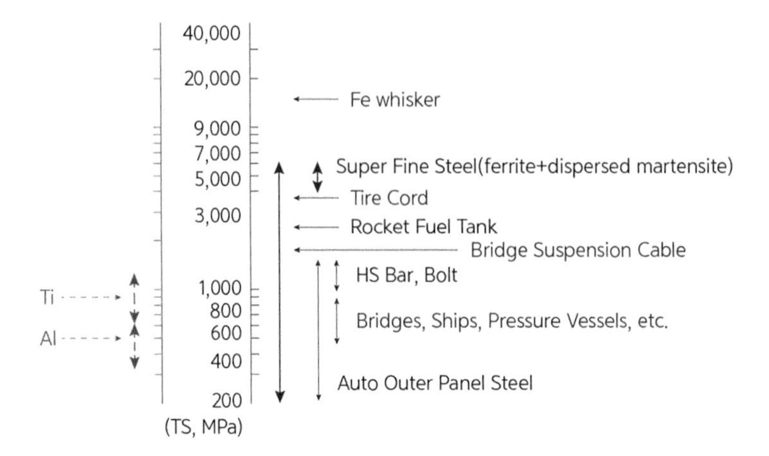

FIGURE 1.12 Tensile strength ranges of various ferrous and common non-ferrous metals.

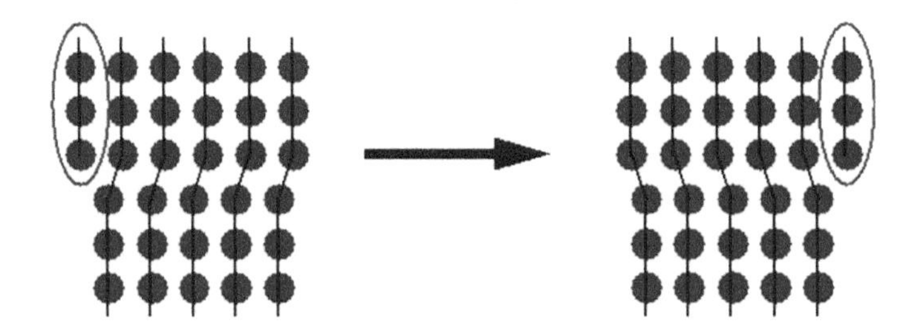

FIGURE 1.13 Representation of a dislocation moving through the crystalline matrix.

that the plastic deformation of a metal takes place when the dislocation moves under the applied load (see Figure 1.13). Therefore, the strength of a crystal increases when the movement of the dislocation is hindered, and the plastic workability increases when the dislocation movement takes place more freely.

There are five types of strengthening mechanisms that are widely used in metallic materials: solid-solution hardening, precipitation hardening, grain refinement hardening, work hardening, and transformation hardening.[(47)] In the case of steel, all of these five strengthening mechanisms can be used in combination. However, this is not the case for non-ferrous metals such as aluminum and copper. In particular, transformation hardening has been the most commonly applied for steel products and the strength improvement is very high, but its contribution to strengthening is meager in aluminum and titanium.

By adding various alloys to iron, the strength can be improved by a mechanism called solid-solution hardening because the alloying elements create a stress field, hindering the movement of dislocations. Precipitation takes place when the amount of an alloying element exceeds the solubility limit of an alloy steel. The amount and type of the precipitate depend on the interfacial properties of the precipitate formed and the thermomechanical conditions. Once a precipitate is formed, a stress field is created around the precipitate, and the movement of dislocation is blocked, resulting in strengthening. The most common precipitates in steel are carbides and nitrides. The impact becomes more prominent as the size of the precipitate becomes finer or the quantity of the precipitate increases.

The grain boundary area is increased by reducing the size of crystal grains. Since the grain boundary plays a role in hindering the movement of dislocations, the strengthening effect becomes more pronounced with finer grain sizes. Grain refinement is widely used in TMCP (thermomechanical control process) steels produced under the careful control of thermal and deformation conditions because it not only increases strength but also improves toughness. Work hardening is a strengthening phenomenon when plastic deformation is applied to a material. This is because many lattice defects such as dislocations form inside the material during plastic deformation and hinder the movement of dislocation.

Transformation hardening occurs when a metal generates a new metastable phase during heat treatment. In steel, the metastable phases formed are mostly bainite and martensite. When steel is rapidly cooled, as in quenching, martensite is mainly formed. Martensite contains numerous dislocations in its microstructure, resulting in high strength and hardness. As a result, martensitic steels are often too brittle and have low toughness. To produce high-strength steel with good toughness, a tempering heat treatment is added after quenching. In the medium cooling range, bainite is formed where bainite is a structure in which cementite is included in various forms in irregularly shaped ferrite. Bainite steel has an intermediate strength between martensitic and ferritic steels. Figure 1.14 shows the microstructure and strength relationship according to the holding temperature when a low carbon steel is treated as follows; heating ⇒ rapid cooling ⇒ holding at a certain temperature ⇒ rapid cooling. It can be seen that the TS can be changed in a wide range from 400 to 1,200 MPa depending on the holding temperature.[48]

The transformation strengthening phenomenon occurs mainly in steel but is rarely observed in other metals. This is because the strengthening

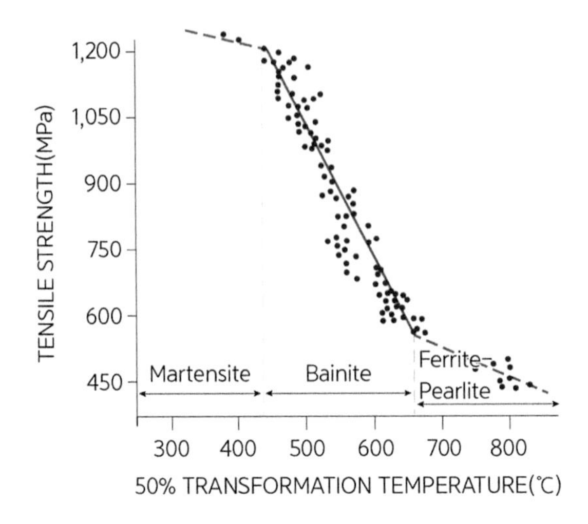

FIGURE 1.14 Variation in the tensile strength of structural steels as a function of the temperature at which the rate of transformation is greatest during continuous cooling heat treatment.[48]

TABLE 1.1 The Behavior of Volume Changes during the Transformation of Various Metals.

Low temp. phase ↔ High temp. phase	Metallic elements
fcc ↔ bcc	Ca, $Ce^{\gamma/\delta}$, $Fe^{\gamma/\delta}$, $Mn^{\gamma/\delta}$, etc.
hcp ↔ bcc	Hf, Li, Na, Sr, Ti, Zr, etc.
bcc ↔ fcc	**$Fe^{\alpha/\gamma}$**
hcp ↔ fcc	Co, etc.
Complex phase transformation	$Mn^{\alpha/\beta/\gamma}$, Sn, $U^{\alpha/\beta/\gamma}$, etc.

is due to the volume expansion of the transformed martensite that occurs during rapid cooling.[49] The behavior of the volume expansion when austenite to cooled to a low temperature has already been shown in Figure 1.9, and is a phenomenon observed only in steel as can be found in Table 1.1. This is the main reason why steel can obtain a high degree of strengthening by transformation. In the case of common non-ferrous metals, the volume contracts when a low-temperature transformation phase is created, which is consistent with the scientific common sense that the volume of all materials decreases when the temperature is lowered. However, in steel, when the high-temperature stable phase austenite is cooled, the low-temperature stable phase ferrite or martensite is formed but the volume increases.[49] This interesting behavior of strengthening in steel is thought to be a gift from God to man along with steel.

1.4.2 Plastic Workability

In order to utilize steel products for various applications, it is necessary to deform sheets or wire rods into finished products of various shapes. To facilitate such plastic deformation, materials with good plastic workability such as ductility and formability are required. Steel is widely used in various industrial fields, including machinery and construction because the material itself is relatively inexpensive and possesses high strength along with excellent unique plastic workability. Steel can be formed into various shapes and is useful in different areas of human life for various purposes. High plastic workability means that the material does not fracture easily and has a large amount of elongation when stress is applied. When a fracture occurs, the material develops cracks on the surface and inside. Fractures can generally be classified into the following four types.[50]

1. Ductile fracture accompanied by plastic deformation.
2. Sudden brittle fracture due to stress increase.
3. Gradual fatigue fracture.
4. Delayed fracture, which occurs when a material is maintained under a static load for a long time.

When a ductile fracture occurs, it is possible to observe to some extent how the fracture will occur because the plastic strain is concentrated around the fracture site. However, in the case of brittle fracture which occurs suddenly, it is difficult to predict because there are hardly any visible symptoms. This sometimes causes significant accidents. An example of brittle fracture is the sudden splitting of a ship anchored in the cold North Atlantic Ocean during World War II.[51]

When an impact is applied to a material, there is a temperature at which the amount of energy that the material can absorb decreases rapidly. This temperature is defined as the ductile-brittle transition temperature. In the case of brittle fracture, the ship's steel plate which has sufficient ductility at room temperature becomes brittle at low temperatures and fractures rapidly. Figure 1.15 shows the trend that as the temperature is lowered the impact energy is decreased for steels with various carbon content.[52] Therefore, if the temperature of the cold sea or air is lower than the transition temperature of the ship's steel plate, even a small impact can cause the steel ship to break.

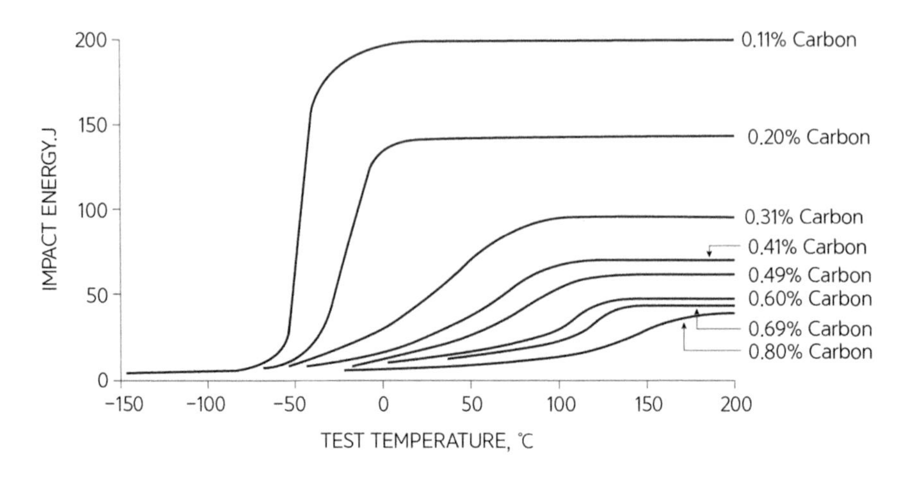

FIGURE 1.15 The behavior of embrittlement according to temperature decreases by the carbon content of the steel.(52)

Fractures observed in mechanical parts are usually fatigue fractures in which stress is repeatedly applied. The fracture occurs in a specific local area, such as a region where stress is concentrated for a geometric reason or a region where there is a defect in the material. A local fracture occurs when the stress in that area exceeds a critical value, which is very difficult to predict. Fatigue fracture is also rarely observed externally, so rigorous preliminary testing and quality control are required in material selection.(50)

Hydrogen embrittlement is a typical delayed fracture phenomenon. When steel is manufactured or used, even a very small amount of hydrogen, on the order of several parts per million, absorbed into the steel sheet can cause brittleness inside the material. Since hydrogen is the smallest atom in size, it can easily diffuse and move even at room temperature. As a result, the migrated hydrogen easily accumulates around voids or defects in the material, forming hydrogen molecules whose volume rapidly increases, increasing the local pressure around the defects, and making them susceptible to rupture.

Aluminum is a very versatile metal although it has lower strength and is more expensive than steel. However, it is soft, light, and has excellent corrosion resistance. Since aluminum is known to be soft, it is sometimes believed that the plastic workability of aluminum is better than that of iron. However, this is not the case. Table 1.2 compares the plastic workability parameters (YS/TS, n, and r) of four steel products with different strength and ductility levels and two aluminum alloys. YS/TS is referred

TABLE 1.2 Comparison of Workability of Four Steels With Different Quality (Hyper-EDDQ, TRIP, TWIP, and STS 304) and Two High-strength Aluminum Alloys (3003 and 6061 grades). (EDDQ: Extra Deep Drawing Quality, TRIP: Transformation Induced Plasticity, TWIP: Twinning-Induced Plasticity.)

Material		YS/TS (MPa)	Yield Ratio (YS/TS)	Uniform El (%)/ Total El (%)	n	r
Steel	Hyper-EDDQ	145 / 280	0.52	25.0 / 54.0	0.27	2.9
	TRIP	400 / 610	0.66	22.0 / 33.0	0.23	1.0
	TWIP	500 / 900	0.56	40.0 / 60.0	-	-
	STS 304	306 / 734	0.42	48.3 / 53.1	0.41	1.1
Al	3003Al	56 / 113	0.50	25.2 / 31.4	0.19	0.75
	6061Al	120 / 221	0.54	21.6 / 23.4	0.22	0.64

to as the yield ratio, and a lower value is desired for good workability. As for n (strain hardening exponent)[53] and r (Lankford coefficient),[54] higher values are desired for good plastic workability. Comparing the values in Table 1.2, the strength of steel is two to six times higher than that of aluminum alloys. The yield ratios of the two alloys are comparable, so it is difficult to say which is superior. However, the data show that the workability parameters such as n and r of steel are consistently higher than those of aluminum, demonstrating the excellent plastic workability of steel which is stronger.

1.4.3 Magnetic Property

Most of the Earth's iron is found in its core. Humans cannot directly use the iron located deep in the core, but iron is critical to human life. The liquid iron contained in the outer core creates a magnetic field around the Earth as it rotates with the Earth's rotation.[55] Although the strength of the Earth's magnetic field is incomparably smaller than the magnetic field of the magnets we use, it is still extremely important to us. It is the Earth's magnetic field that allows us to determine the direction of north and south with a compass made of magnets. Thanks to the Earth's magnetic field we can enjoy the aurora phenomenon, which creates an exquisite spectacle that takes our breath away in the polar regions.

Additionally, the Earth's magnetic field plays a crucially important role in protecting life on our planet. The Earth's magnetic field, formed by iron in the Earth's core as shown in Figure 1.16, serves to block the solar wind and provides a safe environment for life on Earth. The principle of the formation of the Earth's magnetic field is explained by the Dynamo

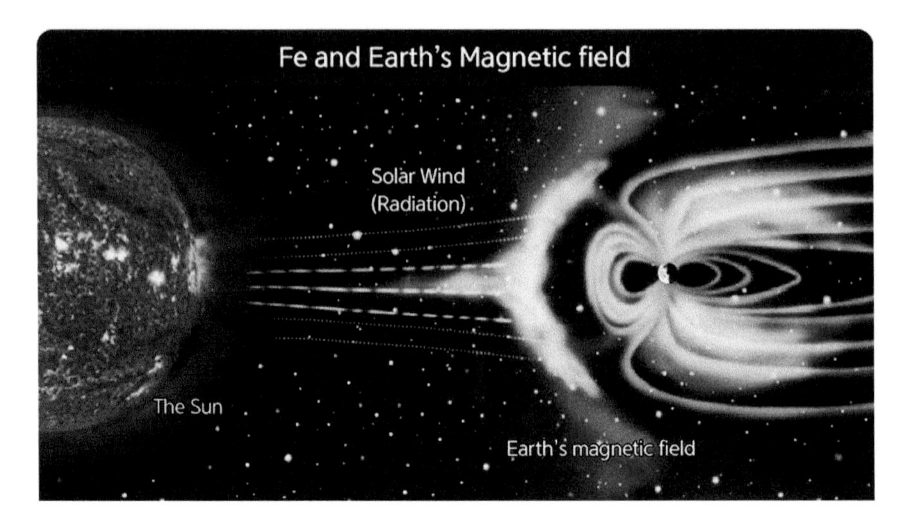

FIGURE 1.16 The role of a shield against the solar wind of the magnetic field around the Earth created by the Dynamo phenomenon.[56] (Courtesy of ESA/NASA, https://www.esa.int/ESA_Multimedia/Images/2007/10/The_Sun-Earth_connection)

theory,[56] which states that the Earth's outer core is made up of electrically conductive liquid iron that readily generates a magnetic field around the Earth. The solar wind refers to the flow of plasma emitted from the upper atmosphere of the Sun, which consists of electrons and protons and is a form of radiation. Exposure to this cosmic radiation causes DNA damage and leads to cancer. Human organs, such as the skin and intestines, can undergo a serious transformation that may eventually lead to death. High-energy particles with electric charge can also damage not only the wireless communication using radio waves but also the power transmission and distribution system of power plants.

In the era of the Fourth Industrial Revolution characterized by hyper-connectivity, super-convergence, and super-intelligence, the internet system is an essential component. If the magnetic field is damaged by the solar wind, the internet could become unusable, leading to significant frustration for humanity. The absence of internet communication would mean no hyper-connectivity and no Fourth Industrial Revolution. In short, iron is the guardian of all organisms on Earth and plays a vital role in enabling human society to create a better future with the internet.

Iron possesses a unique property, magnetism, which allows it to attract magnetic materials. Due to its high magnetization capacity, iron is widely used to make strong magnets. Magnetism is a rare phenomenon that can

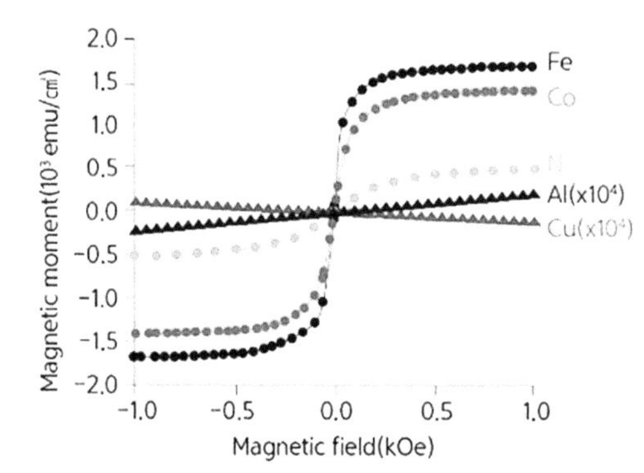

Material	Saturation moment (emu/cm³)
Fe	1,714
78 Permalloy (78Ni, 22Fe)	865
FeCo	1,910
Co	1,422
Ni	484

FIGURE 1.17 Hysteresis curves (left) and saturation magnetic moment of various metals (right).

be observed in a limited number of materials on Earth including nickel and cobalt in addition to iron. However, iron's magnetic properties are superior to those of nickel or cobalt, and hence it is widely used in the modern information age due to its significantly lower cost.(57) In contrast, the price of nickel or cobalt widely used as a secondary battery material is currently skyrocketing.

Figure 1.17 compares the magnetic properties of metals that are widely used as magnetic materials today, that is, magnetic moment and saturation moment. The magnetic moment is a measure of the strength of the magnetic field produced by a magnet. It defines the torque experienced by the magnet in an externally applied magnetic field. It can be observed that the magnetism of iron is superior to that of other metals.(58, 59) Here, the saturation moment refers to the state in which the magnetic moment does not increase anymore as shown in the left picture of Figure 1.17. It is noted that the magnetic moment indicates the degree of magnetism of materials when they enter the magnetic field. By comparing the saturation moment of various materials as shown in Figure 1.17, it can be seen that Fe and Fe alloys have excellent magnetic properties.

It has been a very long time since mankind discovered magnetism and used it in daily life. The compass made using the characteristics of a magnet is referred to as "Chi Nam (司南)" in a book written by Han Fei and others during the Warring States Period in China 3,000 years ago.(60) Chi Nam in Chinese means a compass that shows the four directions on the map. It is said that the use of the compass in Europe also spread from China. In ancient Greece, a natural ore magnet called 'Lodestone' was discovered around the 6th century BC.(61) This one contained a large amount of Fe_3O_4, one of the main iron minerals. It is said to have been named 'Magnetite' because the place of production was "Magnesia."

The Age of Exploration began with the use of natural magnets as compasses. As their usefulness became widely known, mankind placed great emphasis on the development of artificial magnets. However, it was not until the 1900s that artificial magnets were commercially available. Around 1915, iron magnets were produced, followed by KS steel in the 1920s, which added carbon, tungsten, and cobalt to iron, and became commercially available.[(62)] In the 1930s, oxide-based ferrite magnets and Alnico permanent magnets made by melting and casting iron, aluminum, nickel, and cobalt were also developed and widely used.

Rare-earth-based permanent magnets were also developed by adding rare-earth elements such as samarium or neodymium to iron or cobalt. In the 90 years since the development of KS steel, the performance of rare-earth permanent magnets has increased dozens of times, and their use has also greatly expanded. However, these rare-earth permanent magnets have problems with environmental pollution and the weaponization of resources in China. Recently, research has been conducted to develop permanent magnets that do not contain rare earths and have excellent performance.[(63)]

Today, magnets are used for a wide variety of purposes, including microwave ovens, televisions, generators, motors, speakers, and more. It is not an exaggeration to say that the development of human civilization, especially the development of the information industry, has greatly benefited from the effective use of iron and other magnetic materials. In the early 20th century, electrical steel was commercialized. In the middle and late 20th century, magnetic materials were used in various applications such as cassette tapes, computer storage devices, giant particle accelerators, and magnetic levitation trains. They have contributed to the development of modern civilization and emerged as key materials. Magnetism, a mysterious property of iron, will continue to play an important role in the development of human civilization in the present and future.

An electromagnet is a magnet that becomes magnetized when current flows through it and returns to its original state without magnetization when the current is cut off. When current flows through a conductor, a concentric magnetic field is formed around the conductor. This principle can be used to create a very strong magnetic field that cannot be obtained with a permanent magnet. The iron core of the electromagnet is mainly made of soft magnetic material. When the magnetization reaches a certain point, it cannot be increased any further, even if the current is increased. This is known as the magnetic saturation state, which is described in Figure 1.17. An electromagnet can change the strength of a magnetic field by adjusting the current. Therefore, it is widely used from relays in communication devices to electromagnetic cranes that lift heavy materials.

Iron cores are made of electrical steel and are widely used in transformers, generators, and motors, playing a key role in the production, distribution, and utilization of electricity. The most commonly used electrical steel in real life in terms of quantity is Si steel, which is an important material from an industrial point of view. The reason why silicon is most widely used in electrical steel is that it can significantly raise the electrical resistivity of iron while reducing the magnetic anisotropy of iron, thereby reducing the energy loss due to the occurrence of eddy currents.

Electrical steel sheets currently being produced are largely divided into grain-oriented (GO) electrical steel sheets and non-oriented (NO or NGO) electrical steel sheets. GO electrical steel sheets are widely used in stationary devices such as transformers. NO electrical steel sheet is mainly used for large rotating machines or motors that require high-speed rotation, such as electric vehicles. Both GO and NO electrical steel sheets sometimes contain 3% or more Si depending on the application.

In the production of GO electrical steel sheets, the texture should be formed adequately. The texture should be parallel to the rolling direction of the steel sheet, as shown in Figure 1.18 so that magnetization is formed easily.[64] This texture called the Goss texture was first discovered by G. P. Goss.[65] How to obtain this Goss texture well in the process is the key to the GO electrical steel manufacturing technology, and even now, every company protects it as a top-secret treasure.

1.4.4 Weldability

Welding is a fabrication process in which heat or pressure is applied to the contact surface of two metallic materials to bring atoms closer to each other to the point where interatomic forces act. In general, welding is a process in which a portion of the contact surface is melted by a heat source, solidified, and then joined in atomic units. The contact surfaces are mostly joined using a filler metal. Almost all metals can be joined face-to-face by welding, but iron is the metal that can easily secure a sound weld zone. The superior weldability of iron compared to other metals comes from several interesting physicochemical characteristics of iron, such as the melting

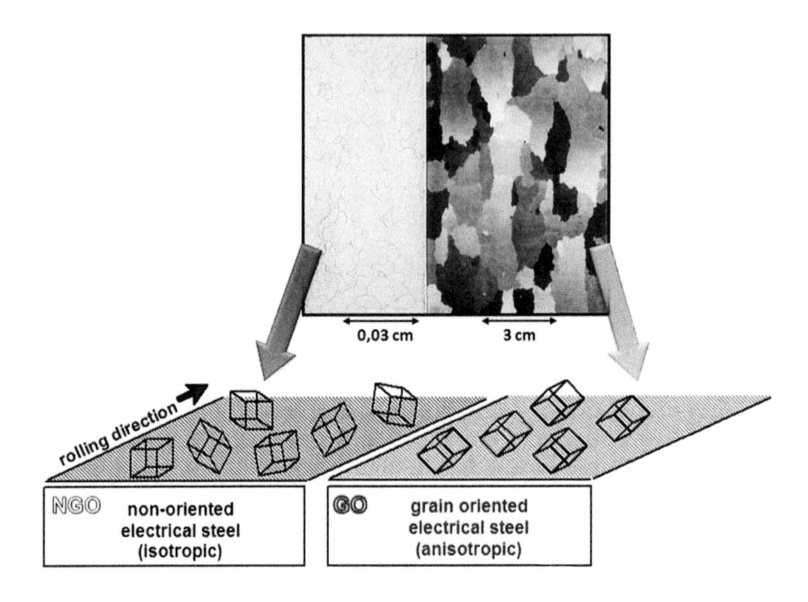

FIGURE 1.18 Random texture (left) and strong crystallographic Goss texture (right) in Fe-Si steels for electrical application. (With permission from Dierk Raabe, https://www.dierk-raabe.com/electrical-steels-fe-3-si/)

point of iron oxide, thermal conductivity, thermal expansion ratio, elastic modulus, and so on.

During welding, the welding metal reacts with oxygen to form oxides because of the high temperatures involved. A notable point is observed when comparing the melting points of the oxides formed with those of various weld metals such as iron, copper, aluminum, magnesium, and titanium as shown in Figure 1.19. The melting temperature of FeO is 1,377°C, which is lower than the melting temperature of iron, which is 1,538°C. In the case of non-ferrous metals such as aluminum, copper, magnesium, and titanium, the oxide formed has a higher melting temperature than the weld metal.[66] If the melting temperature of the oxide is lower than that of the weld metal, such as iron, the oxide exists as a liquid in the weld metal, and it floats to the top rather than affecting the quality of the welded area significantly. For non-ferrous metals, however, the melting temperature of oxides is higher than the melting temperature of the metals. As a result, the solid oxide remains on the welded area even after welding and can penetrate the welding area. External stress can induce a starting point for fracture, making the welded part weak.

Another reason why the weldability of iron is superior to that of non-ferrous metals such as aluminum, magnesium, copper, and titanium is

TABLE 1.3 Comparison of Physical Properties for Weldability Evaluation of Ferrous and Non-ferrous Metals

Metal	Thermal conductivity (W/m•K)	Thermal expansion (x 10^{-6}/K)	Modulus of elasticity (GPa)	Specific heat (cal/g•K)	Yield strength (MPa)
Fe	**80**	**11.8**	**210**	**0.108**	**100–1,500**
Al	237	23.1	70	0.900	20–500
Cu	401	16.5	130	0.093	100–400
Mg	156	24.8	45	0.243	80–250

due to the physical properties of iron. Table 1.3 compares the thermal conductivity, thermal expansion coefficient, elastic modulus. The thermal expansion coefficient of iron are smaller than those of other non-ferrous metals, but the elastic modulus is larger than that of other metals. The specific heat is the highest for aluminum and decreases in the order of magnesium, iron, and copper. Low thermal conductivity means that heat is difficult to transfer to the outside, making it easier to reach the melting temperature required for welding. As a result, the area of the heat-affected zone is small, which is an advantage from a weldability standpoint. The smaller coefficient of thermal expansion reduces distortion in the welded area because the residual stress is reduced during cooling after welding. In addition, if the elastic modulus is large, the residual stress is predicted to be small because the generation of stress due to temperature change is small. Other non-ferrous metals have high thermal conductivity, making it difficult to supply heat to the welded area for joining.

To form a sound welding zone between metallic materials, all relevant factors such as oxidation, metallurgy, and electrical engineering must be considered and optimized. During welding, high temperatures and rapid cooling can lead to stress and deformation, often causing cracks in the weld zone. Such cracks can ultimately lead to the failure of welded structures, so preventing their occurrence during design and construction is a major challenge.

Welding is a complex process involving oxidation, metallurgy, electrical and thermal engineering. Therefore, to improve weldability, it is important to optimize all relevant factors. High-temperature heating during welding can result in residual stress or deformation during cooling, leading to cracking in the weld zone. Among the various defects in the weld zones, cracks are particularly fatal, and preventing their occurrence is a major challenge in designing and constructing welded structures.

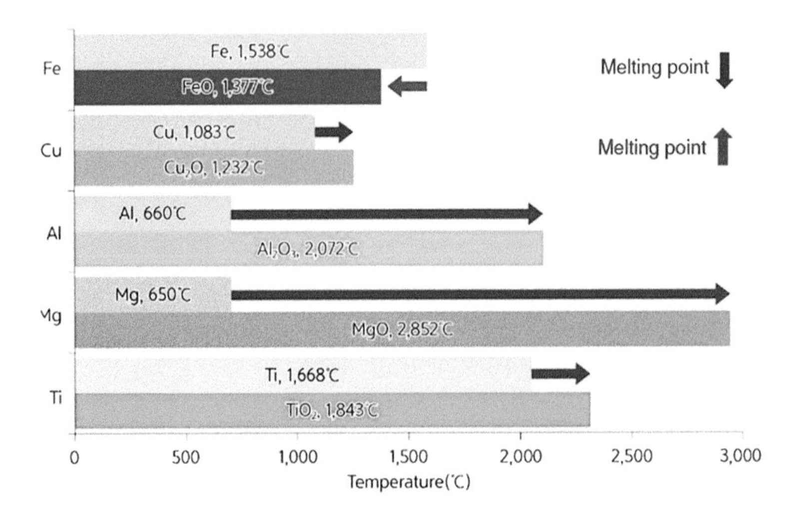

FIGURE 1.19 Melting temperature of weld metal and its oxide formed during welding.

In conclusion, iron has the best weldability among metals, making it the most commonly used metal in the construction of structures of various shapes. Although the economic availability of steel is the primary reason for this trend, excellent weldability is also necessary for the safe construction of structures such as high-rise buildings and super-span bridges. Moreover, welding is essential for realizing structures with diverse and beautiful shapes that artists desire.

There is an anecdote from history that illustrates the excellent weldability of steel. In the early days of World War II, German submarines called U-boats were sinking Allied ships in the Atlantic Ocean, causing significant losses in supplies, and weakening the strength and morale of the US and Britain. However, the situation changed when American engineers developed a new welding technique for shipbuilding that dramatically improved construction speed. They designed ship hulls with standardized modules, which could be welded together at the shipyard, allowing for the efficient construction of ships with the same specifications at different shipyards. The development of underwater arc welding technology further accelerated ship construction. Thanks to the inherent weldability of steel, the US was able to build 2,710 standard-size cargo ships called "Liberty Ships" between 1941 and 1945. These cargo ships played a vital role in transporting military supplies to and from the US and Europe, contributing greatly to the Allied victory in World War II in Europe.[67, 68]

1.5 ECONOMY AND SUSTAINABILITY

Iron is the second most abundant metal after aluminum when limited to the Earth's crust where it can be mined by humans. Since aluminum is produced by electrolysis of molten salt, it consumes a lot of electricity, so the smelting cost is very high. On the other hand, iron can be smelted using coal, which is abundant and relatively cheap. As of 2020, hot-rolled coil, a major steel product, is about US$460 per ton, which is cheaper than water. For example, the price of 1 liter of bottled spring water is about US$1.25 based on the retail price, but when converted to a ton, it reaches US$1,250 per ton, which is about three times the price of steel products of the same weight. The price of aluminum is between US$1,700 and US$2,000 per ton, making steel products less than one-third of that. Furthermore, the specific strength price, which indicates the price per unit of strength, is significantly lower for steel products (US$4.4 ton/MPa) than for aluminum alloy (US$21.2 ton/MPa) and plastic (US$11.3 ton/MPa). This indicates that steel has excellent strength and is cost-effective. While it is difficult to make a direct comparison due to differences in usage and manufacturing processes, it is clear that steel products are significantly less expensive than any other material.

The most common material worldwide is cement, which is mainly used in buildings, with an annual consumption of 2.5 billion tons. Steel is the second most used material after cement, with 1.8 billion tons produced annually. About 300 million tons of plastic are produced annually, and about 40 million tons of aluminum are produced. Copper is used only at 20 million tons per year. Steel is the most preferred building material by architects when constructing residential or commercial buildings.[69]

Durability is a distinct characteristic that demonstrates the economy of steel. Steel can withstand harsh weather conditions such as hurricanes, storms, and even earthquakes. The steel structure of the house is fire-resistant. Steel is completely free from all kinds of weaknesses that wood has, such as insect ingestion, deformation, cracking, splintering, and decay. In addition, steel does not attenuate as quickly as other building materials. Even during building extensions or renovations, steel is inherently flexible enough to be handled without major damage.

Architects discovered early on that steel framing would support longer spans than wooden framing, providing more possibilities for architectural design. Nowadays, architects can create spaces that were not previously possible by utilizing steel products. Steel also has aesthetic qualities

FIGURE 1.20 Inside Terminal #2 of Incheon International Airport showing a wide sense of openness. (With permission from Incheon Airport.)

that architects love to explore. For instance, a steel frame structure can realize a wide indoor space without columns, creating a wonderful sense of openness for people, as shown in Figure 1.20. It is the rigidity of steel that makes the design of stylish building forms possible, giving architects greater freedom to realize their ideas.

Let's enumerate a few reasons why steel is cost-effective.

- Steel is highly fire-resistant, reducing the risk of fire.
- Steel structures are easy to fabricate using welding or fastening methods.
- The quality of steel is proven and reliable.
- Digital modeling techniques can be used to manufacture steel components without defects from the start.
- Because steel is structurally efficient steel-framed buildings are often lighter and require smaller foundations.
- Steel-framed buildings generate less noise, dust, and waste at the construction site than concrete buildings.

- Steel structures can be erected faster with fewer on-site workers, saving on labor costs, equipment rental, and road occupancy costs.
- Construction time is shorter, reducing the opportunity cost of being out of business.

In recent years, smart construction technology has been developed to improve construction efficiency by utilizing the latest information and communication technology (ICT). Smart construction utilizes building information modeling with computer-aided design (CAD) or computer-aided manufacturing (CAM), pre-construction with Virtual Reality (VR) and Augmented Reality (AR), and modular construction with prefabricated steel components. By using these methods, it is possible to shorten the time from design to completion, improve quality, and reduce costs. The modular construction used here is known to be the most efficient when using steel parts. Given that smart construction is the trend of the times, it is believed that the utilization of steel materials in the construction field will greatly increase in the future.[70]

FIGURE 1.21 End-of-life automobiles are recycled as steel scrap. (Courtesy of https://commons.wikimedia.org/wiki/File:Auto_scrapyard_1.jpg)

Traditionally, iron and steel have been symbols of national wealth. In the case of Korea, the share of steel-demanding industries accounted for 36.1% of the gross domestic product in 2010. In addition, the industrial linkage effect is very high. Looking at the forward linkage effect of the steel industry as of 2010, it is 2.12, which is higher than that of petroleum products (1.27), transportation equipment (0.87), and electrical and electronics (1.01). The backward linkage effect such as the raw materials industry is also high, so steel has the highest score at 1.2, along with automobiles at 1.17. As of 2010, the production inducement coefficient, which is the effect of a change in demand for a specific industry's product on the overall production of other industries, was 2.29 for steel, higher than the industry average of 1.95. In other words, when new demand for steel products worth US$8.65 million arises, the production value of the iron-demanding industry increases by about US$20 million.(71)

Steel's high recyclability is another key factor contributing to environmental sustainability. Because steel is durable, products and structures made of steel have a long lifespan. In addition, steel scrap can be easily separated using a magnet. For example, iron and steel components in scrap cars shown in Figure 1.21 are recycled almost 100% by separating them from other materials using magnets. 100% recycling is possible without loss of quality when reproduced using an electric furnace. Recycling steel leads to significant energy and raw material savings. According to data from the World Steel Association (WSA),(72) 1,400 kg of iron ore, 740 kg of coal, and 120 kg of limestone are saved each time one ton of steel scrap is melted to make a new product. Thanks to steel's 100% recyclability, over 22 billion tons of steel have been recycled worldwide since 1900. The production of steel products based on steel scrap accounts for about 25% of global steel production.

Global steel recovery rates by sector are estimated at 85% in construction, 85% in automobiles (nearly 100% in the US), 90% in machinery, and 50% in electrical, and household appliances. The average lifespan of steel products is approximately 35 to 45 years. However, the time to recycling varies, ranging from 15 to 20 years for automobiles and 50 to 200 years for infrastructure and buildings, compared to several weeks for steel packaging including cans. The fact that 90% of the metals used by people around the world is iron is a testament to the excellent cost friendliness.(73) Looking at Figure 1.22, which analyzes the reusability of all metals in the periodic table published in May Nature in 2022, it can be seen that among metals, iron is the second most sustainable material after gold.(74)

SCRAP METAL

In general, metals with a higher end-of-life recycling rate — the percentage of old scraps that is functionally recycled — have lower loss rates.

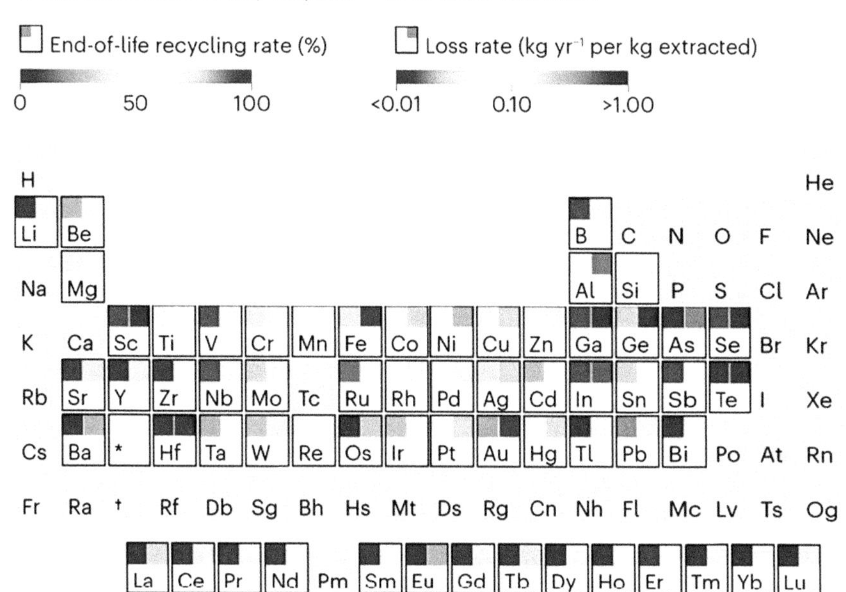

FIGURE 1.22 Recycle rate and loss rate for each metal element.[74] (With permission from Copy Clearance Center.)

REFERENCES

1. John W. Morgan and Edward Anders, "Chemical composition of earth, Venus, and mercury", *Proc. Natl. Acad. Sci., USA.*, Vol. 77, No. 12, December 1980, pp. 6973–6977, doi:10.1073/pnas.77.12.6973, https://www.ncbi.nlm.nih.gov/pmc/articles/PMC350422/
2. Gyana R. Rout and Sunita Sahoo, "Role of iron in plant growth and metabolism", *Rev. Agric. Sci.*, Vol. 3, 2015, pp. 1–24, Released on J-STAGE March 29, 2019, Online ISSN 2187-090X, https://www.jstage.jst.go.jp/article/ras/3/0/3_1/_pdf/-char/en
3. Gene P. Gengelbach, "The importance of micro-minerals: Iron", https://agriking.com/the-importance-of-micro-minerals-iron/
4. "Worldwide prevalence of Anaemia 1993–2005: WHO global database on Anaemia", in edited by Bruno de Benoist, Erin McLean, Ines Egli, and Mary Cogswell, https://apps.who.int/iris/handle/10665/43894

5. "Earth's magnetic field", last edited March 18, 2024, https://en.wikipedia.org/wiki/Earth%27s_magnetic_field
6. "Magnetic and strength properties make these materials essential for electrical products", https://edisontechcenter.org/iron.html
7. "Meteorite", last edited March 24, 2024, https://en.wikipedia.org/wiki/Meteorite
8. "Hoba meteorite", last edited March 20, 2024, https://en.wikipedia.org/wiki/Hoba_meteorite
9. T. Rehren, T. Belgya, A. Jambon, G. Káli, Z. Kasztovszky, Z. Kis, I. Kovács, B. Maróti, Marcos M. Torres, G. Miniaci, Vincent C. Pigott, M. Radivojevic, L. Rosta, L. Szentmiklósi, and Zoltán S. Nagy, "5,000 years old Egyptian iron beads made from hammered meteoritic iron", *J. Archaeol. Sci.*, Vol. 40, 2013, pp. 4785–4792, https://www.sciencedirect.com/science/article/pii/S0305440313002057
10. "Tutankhamun's meteoric iron dagger", last edited March 19, 2024, https://en.wikipedia.org/wiki/Tutankhamun%27s_meteoric_iron_dagger
11. T. A. Wertime, "The beginning of metallurgy: A new look", *Science*, Vol. 182, No. 4115, 1973, pp. 875–887, doi:10.1126/science.182.4115.875
12. Y. Abe Nakai, K. Tantrakarn, S. Omura, and S. Erkut, "Preliminary report on the analysis of an early bronze age iron dagger excavated from Alacahöyük", *AAS*, Vol. 17, pp. 321–324, http://www.jiaa-kaman.org/pdfs/aas_17/AAS_17_Nakai_I_pp_321_324.pdf
13. T. A. Rickard, "The use of meteoric iron", *J. R. Anthropol. Inst. G.B. Irel.*, Vol. 71, No. 1/2, 1941, pp. 55–66. https://doi.org/10.2307/2844401
14. Hideo Akanuma, "The significance of early bronze age iron objects from Kaman-Kalehöyük, Turkey", http://www.jiaa-kaman.org/pdfs/aas_17/AAS_17_Akanuma_H_pp_313_320.pdf
15. R. F. Tylecote, "A History of Metallurgy", 2nd ed., The Institute of Materials, 1992, p. 47, https://www.academia.edu/40301278/HISTORY_OF_METALLURGY_2nd_Edition
16. C. H. Kang, "The Iron Age: The History of Mankind Together with Iron", Gyeonggi-do: Changbi, 2015 (*in Korean*)
17. Y. J. Park, "The history of steel and man", *The Korea Steel Newspaper*, 2002 (in Korean)
18. "75 bible verses about iron", https://bible.knowing-jesus.com/topics/Iron
19. H. R. Schubert, *History of the British Iron and Steel Industry from 450 B.C. to A.D. 1775*, London and Beccles: William Clowes and Sons, Limited, 1955, https://archive.org/details/in.ernet.dli.2015.104051
20. Richard Hayman and Wendy Horton, "Iron Bridge", Gloucestershire: The History Press, 1999
21. Richard J. Fruehan(Editor), "The making, shaping and treating of steel", 11th edition, *The AISE Steel Foundation*, 1998, pp. 475–476.
22. H. C. Allen, *Britain and the United States*, New York: St. Martin's Press, 1955
23. Arthur Herman, "Freedom's Forge: How American Business Produced Victory in World War II", New York: Random House, 2013

24. John Steele Gordon, "An empire of wealth: The epic history of American economic power", New York, London, Toronto, and Toronto: HarperCollins, October 25, 2005, pp. 363–381.
25. Hasegawa Harukiyu, "The Steel Industry in Japan-A Comparison with Britain", Routledge, London and New York, 1996
26. Seiichiro Yonekura, "The Postwar Japanese Iron and Steel Industry: Continuity and Discontinuity." In Etsuo Abe and Yoshitaka Suzuki, eds. *Changing Patterns of International Rivalry: Some Lessons From the Steel Industry. International Conference on Business History 17.* University of Tokyo, 1991, p. 131
27. Tsutomu Kawasaki, "Japan's steel Industry", *Tekko Shimbun Sha*, Tokyo, 1988, pp 30, 213.
28. "POSCO's 50 Years of History 1968–2018", https://www.posco.co.kr/homepage/docs/eng6/jsp/dn/company/posco/Photo_History_of_POSCO.pdf
29. "Steel industry in China", last edited February 2, 2024, https://en.wikipedia.org/wiki/Steel_industry_in_China
30. Rod Beddows, "STEEL 2050-How steel transformed the world and now must transform itself", *Devonian Ventures*, vol. 1-5, 2014
31. "Iron", last edited March 27, 2024, https://en.wikipedia.org/wiki/Iron
32. H. S. Kwon, H. S. Kim, C. J. Park, and H. J. Jang, *Understanding of Stainless Steel*, Korea Iron & Steel Association Stainless Steel Club, 2007, 9-12 (in Korean)
33. "Maraging steel: Properties, processing, and applications", https://www.sciencedirect.com/topics/materials-science/maraging-steel
34. W. Gooch, M. Burkins, D. Mackenzie, and S. Vodenicharov, "Ballistic analysis of Bulgarian electroslag remelted dual hard steel armor plate", 22nd Int. Symposium on Ballistics, Vancouver, BC, Canada, Vol. 2, November 14–18, 2005, pp. 709–716, https://www.researchgate.net/publication/292328380_Ballistic_Analysis_of_Bulgarian_Electroslag_Remelted_Dual_Hard_Steel_Armor_Plate
35. "Shape memory steels", https://www.dierk-raabe.com/shape-memory-steels/
36. Hui Wang, Xiang-Yang Zhou, and Bo Long, "Advanced materials research", Submitted: 2014-07-30, ISSN: 1662-8985, Vol. 1035, pp. 219–224 Accepted: 2014-07-31, doi:10.4028/ www.scientific.net/AMR.1035.219
37. Lasse Lamula, Kari Saarinen, Tomi Lindroos, and Marke Kallio, In Proceedings: 12th International Congress on Sound and Vibration 2005 (ICSV 12). International Institute of Acoustics and Vibration (IIAV), Vol. 6, pp. 4882–4889, https://cris.vtt.fi/en/publications/damping-properties-of-steel-compounds
38. "Heat-Resistant stainless steels – production, alloys, characteristics", https://www.montanstahl.com/blog/heat-resistant-stainless-steels-production-alloys-characteristics/

39. Donald J. Tillack, Joseph E. Guthrie, "Wrought and cast heat resistant stainless steels and nickel alloys for the refining and petrochemical industries", Nickel Development Institute https://nickelinstitute.org/media/1857/wroughtandcastheatresistantstainlesssteelsnickelalloysrefiningpetrochemical_10071_.pdf
40. D. H. Choi, "Manage with Smart Factory", Seoul: Huckleberry Books, 2019, 137–151 (in Korean)
41. "The iron carbon phase diagram", https://www.tf.uni-kiel.de/matwis/amat/iss/kap_6/illustr/s6_1_2.html
42. Paul Gordon, "Principles of Phase Diagrams in Materials Systems", New York: McGraw Hill, 1983
43. William C. Leslie, "The Physical Metallurgy of Steels", New York: McGraw Hill, 1981
44. Taylor Lyman (Editor), "Metals Handbook", 8th ed., Vol. 8, "Metallography, Structures and Phase Diagrams", Cleveland: American Society for Metals, 1972
45. "Iron-Carbon Phases", https://www.metallurgyfordummies.com/iron-carbon-phases.html
46. J. K. Choi, S. G. Lee, Y. H. Park, I. W. Han, and J. W. Morris, "High manganese austenitic steel for cryogenic applications", Paper presented at the Twenty-Second International Offshore and Polar Engineering Conference, Rhodes, Greece, June 2012, https://onepetro.org/ISOPEIOPEC/proceedings-abstract/ISOPE12/All-ISOPE12/13027
47. "What is strengthening and hardening mechanisms of metals – definition", https://material-properties.org/what-is-strengthening-and-hardening-mechanisms-of-metals-definition/
48. K. J. Irvine, F. B. Pickering, F. B. Pickering, W. C. Heselwood, and M. Atkins, "The physical metallurgy of low–carbon, low–alloy steels containing boron", *J. Iron Steel Inst.*, Vol. 186, 1957, p. 54
49. D. Deng, "FEM prediction of welding residual stress and distortion in carbon steel considering phase transformation effects", *Mater. Des.*, Vol. 30, 2009, pp. 359–366.
50. B. D. Youn, "Fatigue Strength and Analysis", School of Mechanical and Aerospace Engineering Seoul National University, https://ocw.snu.ac.kr/sites/default/files/NOTE/7531.pdf
51. D. J. Benac, N. Cherolis, and D. Wood, *J. Fail. Anal. and Preven*, Vol. 16, 2016, pp. 55–66, https://doi.org/10.1007/s11668-015-0052-3
52. T. Armstrong and L. Warner, "Low-temperature transition of normalized carbon-manganese steels", Symposium on Impact Testing. ASTM International, 1956.
53. "Strain hardening exponent", last edited September 26, 2021, https://en.wikipedia.org/wiki/Strain_hardening_exponent
54. "Lankford coefficient", last edited March 20, 2022, https://en.wikipedia.org/wiki/Lankford_coefficient
55. "Earth's magnetic field: Explained", https://www.space.com/earths-magnetic-field-explained

56. "Dynamo theory", last edited February 6, 2024, https://en.wikipedia.org/wiki/Dynamo_theory
57. T. H. Noh, "Magnetic materials", *Duyangsa*, Seoul, 2017, pp. 106–111 (in Korean)
58. "Magnetic moment", last edited March 25, 2024, https://en.wikipedia.org/wiki/Magnetic_moment
59. "Saturation (magnetic)", last edited December 24, 2022, https://en.wikipedia.org/wiki/Saturation_(magnetic)
60. "Compass", https://baike.baidu.com/item/%E5%8F%B8%E5%8D%97/3671419 (in Chinese)
61. "Lodestone", last edited March 25, 2024, https://en.wikipedia.org/wiki/Lodestone
62. "KS_Steel", last edited September 14, 2022, https://en.wikipedia.org/wiki/KS_Steel
63. M. J. Kramer, R. W. McCallum, I. A. Anderson et al., *JOM*, Vol. 64, 2012, pp. 752–763, https://doi.org/10.1007/s11837-012-0351-z
64. "Electrical steels - iron-silicon transformer steels" Raabe, Düsseldorf, Germany, https://www.dierk-raabe.com/electrical-steels-fe-3-si/
65. N. P. Goss, "Electrical sheet and method and apparatus for its manufacture and test", US Patent 1965559, 1934, pp. 1–11.
66. "Welding and joining handbook", in *Ferrous and Nonferrous Materials, The Korea Welding & Joining Society*, Vol. 1, 2008, pp. 35–58, (in Korean)
67. James Davies, "Liberty ship briefing", http://www.ww2ships.com/usa/us-os-001-b.shtml
68. "Building liberty ships for the war effort", 1941, https://rarehistoricalphotos.com/building-liberty-ships-1941/
69. "Material efficiency in clean energy transitions", IEA, March 2019, https://iea.blob.core.windows.net/assets/52cb5782-b6ed-4757-809f-928fd6c3384d/Material_Efficiency_in_Clean_Energy_Transitions.pdf
70. Yishuo Jiang, Ming Li, Daqiang Guo, Wei Wu, Ray Y. Zhong, and George Q. Huang, *Computers in Industry*, Vol. 136, 2022, p. 103594, ISSN 0166-3615, https://doi.org/10.1016/j.compind.2021.103594
71. POSCO's 45th Anniversary, "Beyond steel to become a global No.1 company (2)", *Steel & Metal Newspaper*, Vol. 4, 2013, p. 1 (in Korean), https://www.snmnews.com/news/articleView.html?idxno=305873
72. "Steel and raw materials", Fact sheet from World Steel Association, April 2021
73. "Scrap use in the steel industry", Fact sheet from World Steel Association, May 2021
74. A. Charpentier Poncelet, C. Helbig, P. Loubet, et al. "Losses and lifetimes of metals in the economy", *Nat Sustain*, Vol. 5, 2022, pp. 717–726, https://doi.org/10.1038/s41893-022-00895-8

CHAPTER 2

The Birth of Iron

2.1 THE BIG BANG AND THE FORMATION OF IRON

How was iron in meteorites created? All matter in the universe is composed of 118 elements that appear in Mendeleev's periodic table shown in Figure 2.1. All elements were created during the process of the Big Bang when the universe was formed.[1] However, why does iron exist on Earth more than any other element? Let's understand the origin of the elements and explore why iron differs from other elements and where it came from.

To answer questions about the creation of the elements that make up the periodic table, we need to go back to the Big Bang when the creation of the universe began (see Figure 2.2). The Big Bang happened 13.8 billion years ago when energy was densely packed into a very small point and began to transform into matter as it expanded after a massive explosion.[2–6] After the Big Bang, the following are the main points related to the creation of the elements, and the element iron has a close relationship to the birth of the solar system, i.e., the Earth.

- Universe Age ~ 1 Second: Inflation and the Birth of Subatomic Particles (Quarks, Electrons, Protons, Neutrons, Mesons, etc.).
- Universe Age ~ 3 Minutes: Nucleosynthesis and the Production of Hydrogen and Helium.
- Universe Age ~ 400 Million Years: Formation of Nebulae, First Star Formation, and the Production of Elements Lighter than Iron.

DOI: 10.1201/9781003419259-2

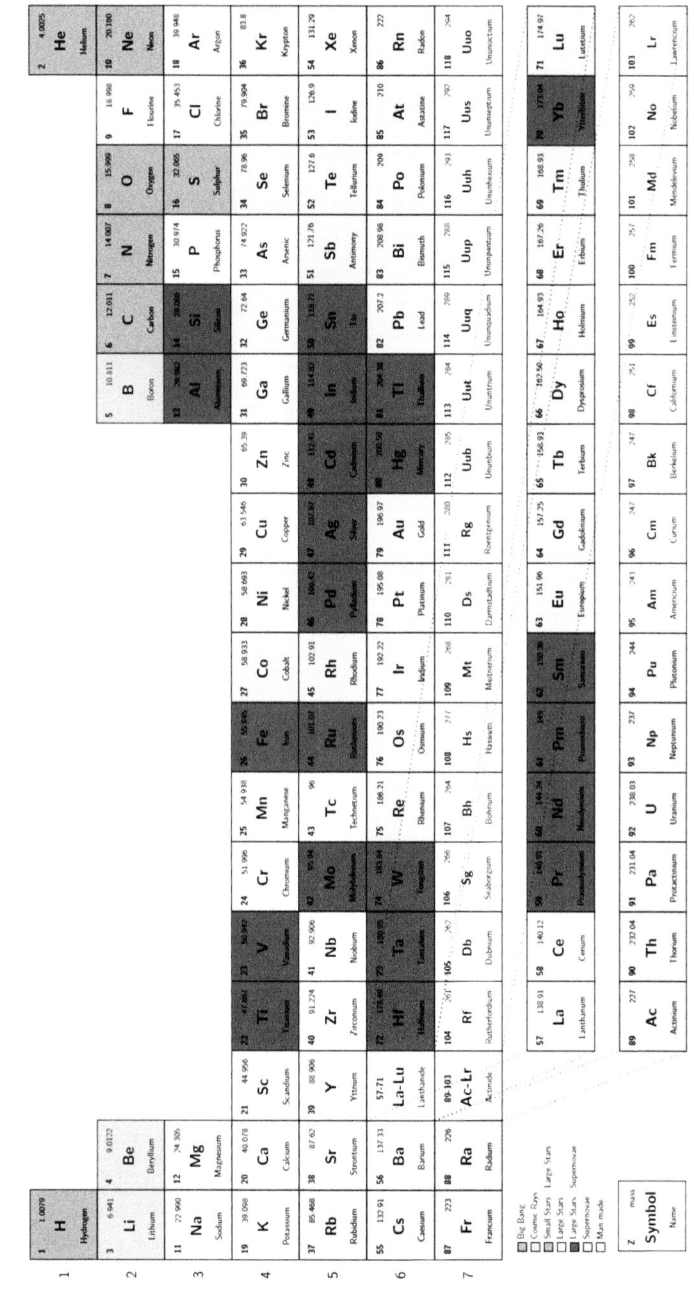

1	1 1.0079 H Hydrogen																	2 4.0025 He Helium
2	3 6.941 Li Lithium	4 9.0122 Be Beryllium											5 10.811 B Boron	6 12.011 C Carbon	7 14.007 N Nitrogen	8 15.999 O Oxygen	9 18.998 F Flourine	10 20.180 Ne Neon
3	11 22.990 Na Sodium	12 24.305 Mg Magnesium											13 26.982 Al Aluminium	14 28.086 Si Silicon	15 30.974 P Phosphorus	16 32.065 S Sulphur	17 35.453 Cl Chlorine	18 39.948 Ar Argon
4	19 39.098 K Potassium	20 40.078 Ca Calcium	21 44.956 Sc Scandium	22 47.867 Ti Titanium	23 50.942 V Vanadium	24 51.996 Cr Chromium	25 54.938 Mn Manganese	26 55.845 Fe Iron	27 58.933 Co Cobalt	28 58.693 Ni Nickel	29 63.546 Cu Copper	30 65.39 Zn Zinc	31 69.723 Ga Gallium	32 72.64 Ge Germanium	33 74.922 As Arsenic	34 78.96 Se Selenium	35 79.904 Br Bromine	36 83.8 Kr Krypton
5	37 85.468 Rb Rubidium	38 87.62 Sr Strontium	39 88.906 Y Yttrium	40 91.224 Zr Zirconium	41 92.906 Nb Niobium	42 95.94 Mo Molybdenum	43 96 Tc Technetium	44 101.07 Ru Ruthenium	45 102.91 Rh Rhodium	46 106.42 Pd Palladium	47 107.87 Ag Silver	48 112.41 Cd Cadmium	49 114.82 In Indium	50 118.71 Sn Tin	51 121.76 Sb Antimony	52 127.6 Te Tellurium	53 126.9 I Iodine	54 131.29 Xe Xenon
6	55 132.91 Cs Caesium	56 137.33 Ba Barium	57-71 La-Lu Lanthanide	72 178.49 Hf Hafnium	73 180.95 Ta Tantalum	74 183.84 W Tungsten	75 186.21 Re Rhenium	76 190.23 Os Osmium	77 192.22 Ir Iridium	78 195.08 Pt Platinum	79 196.97 Au Gold	80 200.59 Hg Mercury	81 204.38 Tl Thallium	82 207.2 Pb Lead	83 208.98 Bi Bismuth	84 209 Po Polonium	85 210 At Astatine	86 222 Rn Radon
7	87 223 Fr Francium	88 226 Ra Radium	89-103 Ac-Lr Actinide	104 261 Rf Rutherfordium	105 262 Db Dubnium	106 266 Sg Seaborgium	107 264 Bh Bohrium	108 277 Hs Hassium	109 268 Mt Meitnerium	110 281 Ds Darmstadtium	111 280 Rg Roentgenium	112 285 Uub Ununbium	113 284 Uut Ununtrium	114 289 Uuq Ununquadium	115 288 Uup Ununpentium	116 293 Uuh Ununhexium	117 292 Uus Ununseptium	118 294 Uuo Ununoctium
				57 138.91 La Lanthanum	58 140.12 Ce Cerium	59 140.91 Pr Praseodymium	60 144.24 Nd Neodymium	61 145 Pm Promethium	62 150.36 Sm Samarium	63 151.96 Eu Europium	64 157.25 Gd Gadolinium	65 158.93 Tb Terbium	66 162.50 Dy Dysprosium	67 164.93 Ho Holmium	68 167.26 Er Erbium	69 168.93 Tm Thulium	70 173.04 Yb Ytterbium	71 174.97 Lu Lutetium
				89 227 Ac Actinium	90 232.04 Th Thorium	91 231.04 Pa Protactinium	92 238.03 U Uranium	93 237 Np Neptunium	94 244 Pu Plutonium	95 243 Am Americium	96 247 Cm Curium	97 247 Bk Berkelium	98 251 Cf Californium	99 252 Es Einsteinium	100 257 Fm Fermium	101 258 Md Mendelevium	102 259 No Nobelium	103 262 Lr Lawrencium

Big Bang
Cosmic Rays
Small Stars Large Stars
Large Stars
Large Stars Supernovae
Supernovae
Man made

Z mass
Symbol
Name

FIGURE 2.1 Periodic table with origins of elements.[1] (Courtesy of Figshare.com, https://figshare.com/articles/figure/Periodic_Table_of_Nucleosynthesis/1595995/1)

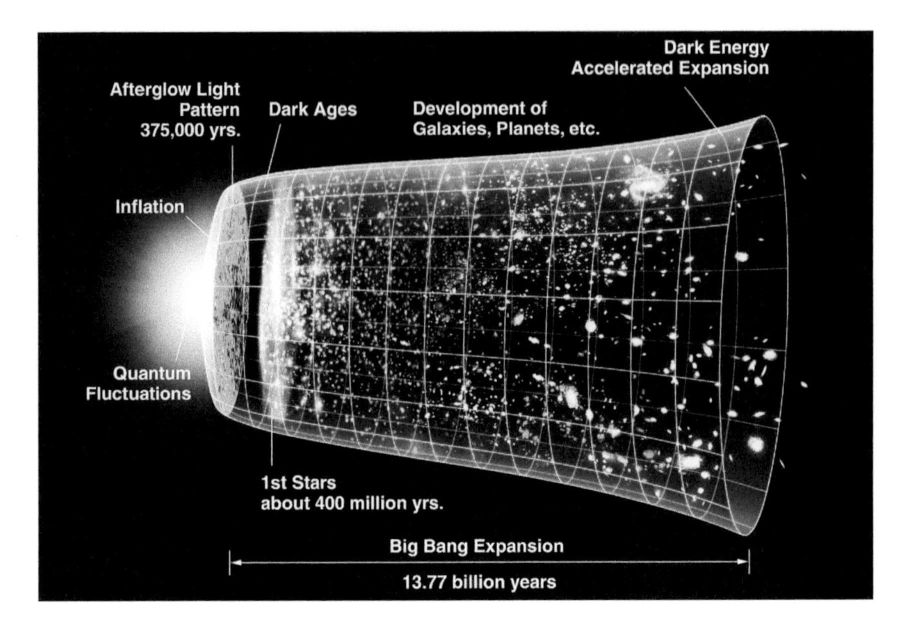

FIGURE 2.2 The Big Bang, the origin of the universe (NASA).[3] (Courtesy of https://commons.wikimedia.org/wiki/File:Inflation_Universe.png.)

- Universe Age ~ 9 Billion Years: Supernova Explosions, Birth of the Solar System, and the Production of Elements Heavier than Iron.
- Universe Age ~ 13.8 Billion Years: Current Universe.

It has been demonstrated by Einstein's theory of relativity that the energy that existed in the early universe transformed into all the elements that made up the universe. Einstein's special theory of relativity, announced in 1905, is expressed in the following equation: in this equation, E is energy, m is mass, and c is a constant that represents the speed of light.

$$E = mc^2 \tag{2.1}$$

Einstein's theory of special relativity, which stated that energy and matter can be converted into each other, was a moment when the law of conservation of energy, known as the foundation of thermodynamics, established by Lavoisier (1743–1794), collapsed. This theory provided the theoretical basis for atomic bombs, hydrogen bombs, and nuclear power generation, and explained how stars like the Sun can continue to generate so much energy through nuclear fusion. With the Big Bang explosion, the

◀ H = 1P + 1E
◀ He = 2P + 2N + 2E
◀ Li = 3P + 3N +3E

• •

• •

◈ Above elements were formed by nuclear fusion!

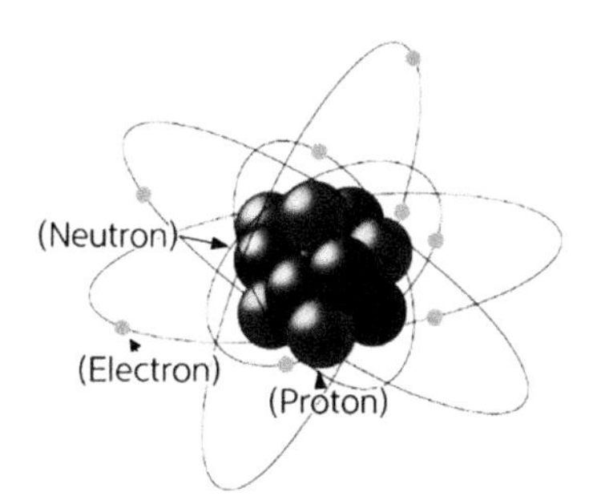

FIGURE 2.3 Structure of an atom composed of P – Proton, N – Neutron, and E – Electron. (H: Hydrogen, He: Helium, Li: Lithium)

universe expanded, and during this expansion process, quarks fused to create elements as shown in Figure 2.3.[7]

Looking at the internal structure of an element, there is a nucleus in the center with electrons orbiting around it, and the nucleus exists in the form of a combination of protons and neutrons.

Hydrogen, which has an atomic number of 1, consists of one atomic nucleus containing a single proton and neutron, and one electron rotating around the nucleus. Helium, with an atomic number of 2, consists of one atomic nucleus containing two protons and neutrons, and two electrons rotating around the nucleus. The atomic number of an element is determined to match the number of protons within the element, and each element in the periodic table has a unique atomic number that increases as the number of protons, neutrons, and electrons increases accordingly.

Scientists have discovered that there are four fundamental forces, namely, gravity, the electromagnetic force, the strong force, and the weak force, that act among protons, neutrons, and electrons that make up elements. Gravity causes objects to fall to the ground and makes the Moon revolve around the Earth, and it also plays an important role in the giant close-range explosion that occurs when stars contract and disappear, transforming into supernovae. The electromagnetic force is a force that binds atomic nuclei and electrons together to form atoms or binds atoms and atoms together to form molecules or crystals. A strong force and weak force are relatively unfamiliar forces discovered later than gravity and electromagnetic forces, and they are forces that act on a very small scale within atomic nuclei. The strong force binds quarks inside protons and neutrons while strongly binding protons and neutrons together in atomic nuclei. The weak force is a very weak force, about 10^{-12} times weaker than the electromagnetic force, that acts to cause elements such as uranium and radium to undergo radioactive decay. Ultimately, the stability of an element can be said to be determined by the balance of interaction between the subatomic particles that make up the element.

Hydrogen is the first and most abundant element that formed after the Big Bang. Through the process of nuclear fusion in which two protons and two neutrons combine, helium is produced. While heavier elements than helium can be created through this continuous process of nuclear fusion, eventually the reaction stops due to reasons such as temperature and pressure drops resulting from the expansion of the universe.

The composition of the elements in the universe and how much of each element exists can be determined by analyzing starlight using a spectrometer. According to the analysis results, hydrogen is the most abundant element in the universe, accounting for more than 90% of all elements by number and 75% by mass. The next most abundant element is helium, which accounts for 7.1% of all elements by number and 23% by mass. All other elements combined make up only 0.1% by number and 2% by mass.[8] Hydrogen and helium are mostly formed within three minutes after the Big Bang. Due to the rapid expansion and cooling, conditions are not suitable for the synthesis of heavier elements. Heavier elements are produced during the formation and destruction of stars, as explained later.

George Gamow (1904–1968) and Ralph Asher Alpher (1921–2007) developed a model that predicts how hydrogen and helium are formed within a few minutes after the Big Bang, as well as the ratio in which they are created. The model showed remarkable similarity to observed phenomena, leading to significant scientific advancement.[9]

After the Big Bang, the universe consisted mainly of light elements such as hydrogen and helium formed by nuclear fusion, with a uniform distribution throughout the universe initially. However, when temperature and pressure imbalances occurred in some parts of the universe, equilibrium was disrupted, and gravity caused the formation of compact, solid objects called stars. As the conglomerates grew larger, gravity increased, and the coalescence accelerated. As this process repeated countless times, atoms in the center of the star formed heavier atoms through nuclear fusion reactions, producing heat. When the temperature exceeded 10 million degrees Celsius due to the heat generated, the star exploded on a massive scale and ended its life. From around 200 million years after the Big Bang, the universe was filled with numerous stars, which gathered to form galaxies. It is said that there are over 100 billion such galaxies in the universe.

The lifespan of stars depends on their mass. Small stars with less than half the mass of the Sun gradually shrink over billions of years by burning their internal hydrogen. However, large stars with masses ranging from half to eight times that of the Sun generate helium after consuming all of their hydrogens creating high temperatures in the core of up to 200 million degrees Celsius, unlike small stars. When the hydrogen fuel runs out, stars begin to burn helium in their cores to maintain high temperatures. Helium atoms can fuse intact to create even-numbered elements, and sometimes fusion occurs when protons and neutrons decompose, resulting in the creation of odd-numbered elements. As nuclear fusion reactions progress, elements such as lithium, boron, beryllium, and carbon accumulate inside stars. The energy generated by burning helium is less than that generated by burning hydrogen. As a result, most stars run out of fuel in hundreds of millions of years and, at this stage, after passing through the stages of red giant and white dwarf stars, they disappear at the end of their lifetimes.

Heavy stars with masses greater than eight times that of the Sun continue to sustain the next stage of nuclear fusion using carbon as fuel. In this process, elements such as nitrogen, oxygen, fluorine, neon, sodium, and magnesium are created while nuclear fusion energy is supplied, raising the temperature to over 1 billion degrees Celsius. Many stars disappear at this stage due to the end of their lifespan, but some very heavy (more than ten times the mass of the Sun) and hot stars (with an internal temperature of 5 billion degrees Celsius) continue to burn the produced elements through nuclear fusion and generate heavier elements such as silicon, calcium, manganese, iron, and nickel, which are heavier than magnesium while maintaining their energy output for millions of years. At this point, the temperature of the star reaches over 3 billion degrees Celsius, and ultimately, only very large stars are capable of producing heavy elements such as iron. In the famous 1957 B2FH (Geoffrey Burbidge, Margaret Burbidge, William Fowler, Fred Hoyler) paper,[(10)] which tracks the various nuclear fusion reactions in detail, the process of creating all elements up to iron within the star is explained, revealing much of the mystery surrounding element creation.

In cases where nuclear fusion reactions occur continuously in stars more than ten times the mass of the Sun, iron is eventually created because the atomic nucleus structure of iron is the most stable. Scientists have theoretically calculated the binding energy between nucleons that make up atomic nuclei, and the results are shown in Figure 2.4. The binding energy

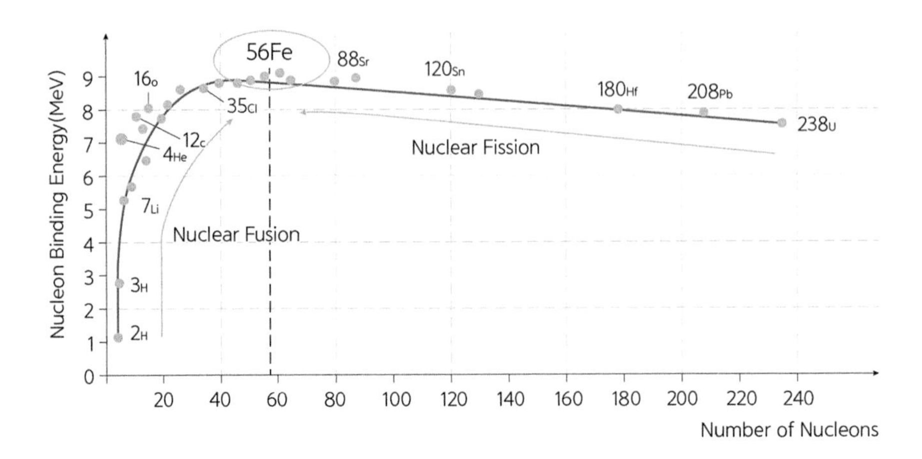

FIGURE 2.4 Nucleon binding energy and nuclear reactions of elements.

between the nuclei gradually increases as the atomic number increases, but it reaches a maximum at iron with atomic number 26 and then gradually decreases. The fact that the binding energy between iron nuclei is the highest means that iron is the most stable.

If the weight of a star is large enough and nuclear fusion continues to occur, elements heavier than iron can also be created. When elements are heavier, they tend to return to iron because they are less stable than iron. In this case, elements heavier than iron can be converted back to iron through nuclear fission. Therefore, it can be said that the likelihood of the existence of iron is relatively higher in stars that are more than ten times the mass of the Sun. However, in terms of the entire universe, the amount of iron is very small compared to hydrogen or helium.

Whether it is a nuclear fusion reaction or a nuclear fission reaction, the reaction can only proceed if external conditions such as temperature and pressure are satisfied. After the Big Bang, as the universe expands, the temperature decreases and the pressure decreases, so the conditions for nuclear reactions may not be met, and under these conditions, the reaction stops, and no more elements are created. In the final stage, the heavy elements created are collected in the middle when a star is formed, and around them, lighter elements are formed, creating a layered structure of the star (see Figure 2.5). The number of layers of elements and how they overlap in a layered structure star depends on the mass of the star, but in the central part, it shows a structure in which iron is mostly present, and layers of elements lighter than iron are created on the periphery. This is also why the interior of the Earth is mostly composed of iron.

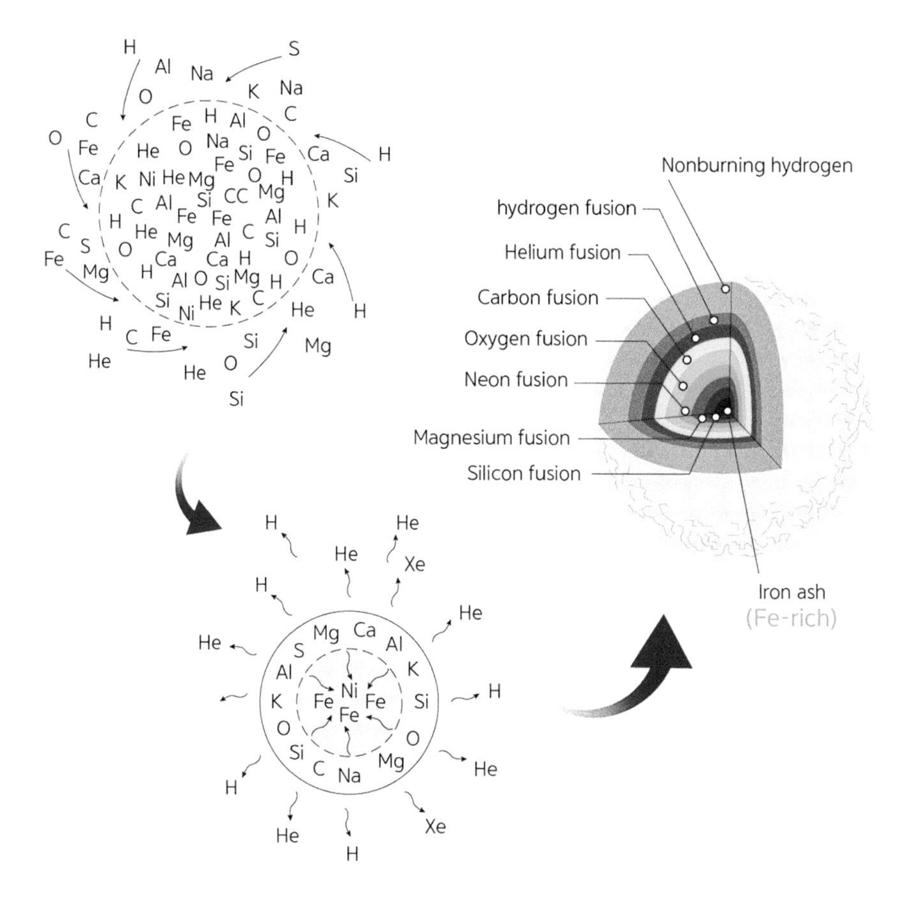

FIGURE 2.5 Star evolution by internal nuclear reactions, counterclockwise from initial formation → contraction → onion structure.

Of the 118 elements listed in the periodic table, only 98 elements exist in nature.(11) In the previous section on the periodic table we mentioned that in the first stage after the Big Bang, 26 elements, up to iron, were created through nuclear reactions inside stars. But how are the elements heavier than iron created? These elements are believed to have been formed by the explosion of a giant star known as a supernova, which is thought to exist in the current galaxy.

Supernovae are a phenomenon in which a star, at the end of its life, collapses due to its own gravity, emits a large amount of light, and enters a destructive stage. The stars in the universe are maintained by the balance between gravity, which causes contraction due to their mass, and the expansion force created by particles produced by nuclear fusion reactions. When the nuclear fusion reaction decreases, the equilibrium is broken, and as a result of the tremendous force created by the gravitational contraction, a strong shock wave is formed, causing the star to decompose. This phenomenon is called gravitational collapse. In a supernova where gravitational collapse occurs, the star dies with an explosion.

In the final stage of a star's evolution, its diameter can increase to several times or even thousands of times that of the Sun, and it can become a cool red giant with a low surface temperature. At this stage, helium is converted into carbon and oxygen atoms, and these atoms begin to transition into heavy atoms such as iron through nuclear fusion. Once iron atoms are produced in the star's core, it becomes difficult to create heavier elements through nuclear fusion reactions, and the star's gravity causes it to contract.

When this star reaches the limit of its contraction and becomes a solid mass, the surrounding material falls in due to gravity, colliding with the solid iron core and increasing its mass, creating a shock wave. When the star's mass reaches a certain limit (1.4 times the mass of the Sun), nuclear reactions occur explosively, causing atomic nuclei to break apart, transitioning the star to the supernova stage and ending its life cycle.

To dismantle the atomic nucleus of an element created by nuclear fusion, a significant amount of external energy must be supplied. More energy is required to dismantle the atomic nucleus of iron created at the center of a star than to dismantle the nucleus of any other element. Since the star is no longer receiving energy from nuclear fusion reactions, it begins to collapse rapidly due to gravity. As the atoms within the star break apart, the temperature of the nucleus rises to over 100 billion degrees Celsius, supplying energy.

When a gravitational collapse occurs, the tension between the atomic nuclei overcomes gravity, and the atomic nuclei are bounced out of the star's core in a shock wave. This is what we see as a supernova explosion (Figure 2.6). During this shock, the star meets atoms in its outer shell, which become heated and undergo rapid nuclear fusion reactions, creating new elements and radioactive isotopes heavier than iron. While elements lighter than iron are mostly created through nuclear fusion between elements, the creation of elements heavier than iron requires the unstable premise of a supernova explosion.

The material created by the shockwave of a supernova expands into space and forms a nebula. The material that explodes far away from a star is known as supernova remnants, which contain lighter and heavier elements than iron. When a massive explosion occurs, resulting in the

FIGURE 2.6 The Crab Nebula is a six-light-year-wide expanding remnant of a star's supernova explosion. (Courtesy of https://en.wikipedia.org/wiki/File:Crab_Nebula.jpg.)

end of a star's life as a supernova, iron atoms are decomposed along with oxygen and carbon atoms and are then expelled far and wide into space. The explosion of a supernova, which emits an enormous amount of light, releases vast amounts of material into space, which eventually becomes interstellar matter, the building blocks for new stars.

After the elemental gases that were emitted during a supernova explosion have disappeared, the remaining atomic nuclei and electrons combine to form a very heavy but very small "neutron star." Stars that are more than three times the mass of the Sun undergo infinite contraction after an explosion, sometimes creating black holes. Stars made from interstellar matter scattered throughout the universe by supernova explosions go through the process of becoming main sequence stars in their adolescence, then pass through red giants, white dwarfs, neutron stars, or black holes before reaching the end of their lives.

When we look up at the black night sky, the universe appears to be static and unmoving. However, the universe's static nature can be shattered by tremendous explosions, such as when stars become supernovae and emit immense amounts of light. Supernovae depict the death of stars, but ironically, they are called "new stars" because they appear to be new for a while to the naked eye before gradually fading away over several months.

Scientists have been observing unprecedented supernovae. One team of scientists reported a new type of supernova explosion on June 17, 2018, in Hawaii and presented it to the academic community.[(12)] However, they have not yet provided a clear explanation for the phenomenon. These scientists reported that the massive flash occurred in a galaxy 200 million light-years away from Earth and was 10–100 times brighter than a typical supernova.

Scientists initially believed that the burst of light named "Cow" originated from our galaxy, but later observations using a spectrograph revealed that the explosion occurred in another galaxy located in the direction of the constellation Hercules. What surprised scientists even more than the brightness was the fact that most supernovae reached their peak brightness several weeks after the initial explosion, but Cow reached its peak brightness in just two days. The astronomical community is now researching to determine whether this is a new type of supernova or something more exotic.

Supernovae are rare events statistically, occurring only a few times per century throughout our entire galaxy. However, supernovae are incredibly important as without them, there would not be enough energy over sufficient periods to create elements such as copper, gold, lead, and uranium that are heavier than iron. When specific fragments produced by the supernova explosion of a nebula containing these elements travel long distances through space and encounter other celestial bodies, they become meteorites.

2.2 FORMATION OF THE EARTH AND IRON

All stars and planets in the solar system were formed by the interaction of interstellar clouds, combining, breaking apart, expanding, and contracting. Earth is a planet within the solar system composed of a star called the Sun. The solar system, to which Earth belongs, is a post-starburst system that was formed relatively late after a supernova explosion. Therefore, many planets such as Mercury, Venus, Jupiter, and Earth in the solar system contain heavy elements such as iron, nickel, and uranium. The origin of the Earth is the same as that of the Sun. The theory of solar system formation has evolved over centuries, but it was not until the 18th century that it took on the framework of modern theory.

With the opening of the space age in the 1950s and the discovery of exoplanets in earnest since the mid-1990s, existing theories about the creation and demise of the solar system have been challenged and refined. According to the currently known theory, the solar system was formed about 4.6 billion years ago as part of a giant molecular cloud that collapsed under gravity. Most of the collapsed mass was concentrated in the center, forming the Sun, and the remaining mass was scattered into planets, satellites, and asteroids, evolving into a disk-shaped primitive planetary system. This hypothesis is the interstellar cloud model theory of the formation of the solar system.[13]

The central part of the solar system collapsed due to its gravity and evolved from a primordial Sun to the present-day Sun. And as particles in the solar nebula disk rotate and condense, small planetesimals are created as the density increases. It is estimated that about 10 trillion of these small planetesimals were produced in the primitive solar system. These protoplanets collide and merge or break apart, and as the collision energy is absorbed, the temperature rises. As the temperature inside the protoplanet increases due to the kinetic energy from the collisions, nuclear reactions of the elements created during a supernova explosion occur, generating additional heat and causing the protoplanet to become molten and extremely hot.

The nuclear reactions that occur in this state proceed in the direction of increasing the iron content, as discussed in the previous section (refer to Figure 2.5). On the surface of the planetesimal, all substances are gaseous due to superheating, forming a primitive atmosphere, and when the primitive Earth cools down, water vapor condenses and precipitates, forming the ocean. In terms of the mass of the Earth, iron accounts for the highest proportion at 32%, followed by oxygen (30%), silicon (15%), magnesium (14%), and sulfur (2.9%). In the lithosphere, which is the Earth's crust, iron accounts for the fourth highest proportion at 5.2%, following oxygen (46%), silicon (28%), and aluminum (8.3%).[14] The Sun accounts for 99.9% of the total mass of the solar system, with most of it being hydrogen and helium. Why is there a significant difference in the proportions of elements on Earth compared to those in the Sun or other stars? Figure 2.7 shows the internal structure of the Earth. Although it has a layered structure similar in concept to that of a star with an onion-like layered structure shown in Figure 2.5, it is not the same. What is the reason for this?

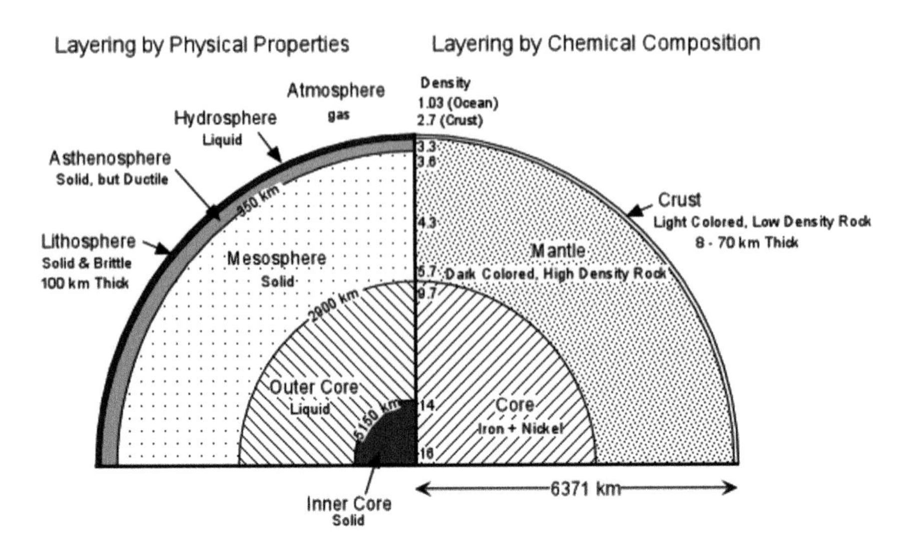

FIGURE 2.7 The internal structure of the Earth. (With permission from Stephen A. Nelson, https://www2.tulane.edu/~sanelson/Natural_Disasters/struct&materials.htm.)

In the stars seen in Figure 2.5, only elements with atomic numbers lower than iron and nickel are present. In contrast, on the Earth, shown in Figure 2.7, all naturally occurring elements listed in the periodic table exist. This fact is due to the formation of the Earth after a supernova explosion. Elements with atomic numbers higher than iron and nickel can only be created by supernova explosions. Supernovae are infrequently occurring and are only observed a few times in a century, meaning that there are only very small amounts of heavy elements in the universe when viewed as a whole. It also means that only a very small number of stars in the universe are formed after experiencing a supernova. Therefore, most stars are thought to be composed mainly of hydrogen and helium, with only a few containing heavier than helium.

Another difference between Figure 2.5 and Figure 2.7 is that while multiple layers are visible in the star in Figure 2.5, the Earth is composed of four major layers: the crust, mantle, outer core, and inner core. The outer core, located within the mantle, is believed to be a liquid sphere with a radius of about 3,500 km, while the inner core is thought to be a solid sphere with a radius of 1,200 km. Although direct exploration is impossible due to the extreme depth, elements can be indirectly identified by seismic waves passing through them, and it is known that 85% of the core's composition is iron and the remaining 10% is nickel. The reason why the inner core is solid is that the melting point increases as the pressure inside it increases.

The mantle, which surrounds the core, accounts for 82% of the Earth's volume and 68% of its mass. It is believed to be composed of rocks containing minerals such as iron and silicon. The melting point of the rocks is higher than that of iron, so as the primitive Earth in a liquid state cooled from the outside, materials with a higher melting point solidified first and took their place. The outermost layer of the mantle is a plate structure, and materials dissolved in the mantle are intermittently expelled to the surface, resulting in volcanic activity. The crust consists of continental and oceanic rocks containing various minerals, and iron is concentrated in certain locations within the crust in the form of ore veins.

The fact that iron exists as a vein in a particular location can be deduced from understanding the process of the Earth's formation. When the Earth was first formed, about 3.8 billion years ago, and primitive oceans were formed, iron that was uniformly distributed on the Earth's surface flowed into the primitive ocean due to rain and sedimentation occurring as this iron gathered in certain areas. At that time, there was not enough oxygen in the Earth's atmosphere to create iron oxide, and iron existed in a pure state, but it became corroded as it was washed away by rainwater and seawater. Rainwater and seawater contain a lot of carbon dioxide or sulfur, making their acidity very high, which could dissolve iron.(15, 16)

After that, microorganisms such as primitive cyanobacteria were created on the Earth and carbon metabolism became active, resulting in a large amount of oxygen being produced. This oxygen reacted with iron to form mineral veins composed of iron oxides or hydroxides that were deposited beneath the sea. Microbial activity varies in time and season, and the degree of oxygen production is different, resulting in different types of minerals and stratigraphic structures. Iron ore veins are created when the strata of the sea floor are raised by plate movement and seismic activity. Humans who discovered these veins could extract iron ore and refine iron through technological advancements spanning thousands of years, making it a useful material for everything on Earth, and by utilizing this iron, humans have become the lord of all things on Earth.

The distribution of iron ore found on Earth is not uniform. According to a recent survey, crude iron ore is estimated at 180 billion metric tons.(17) 51, 34, 20, and 29 billion metric tons of crude iron ore are distributed in

Australia, Brazil, China, and Russia, respectively, and these four countries account for 74% of the world's reserves. About 2,600 million metric tons of ores are mined annually.

The atomic number of iron is 26, and its atomic weight is 55.85. The element symbol of iron "Fe" originates from the Latin word "Ferrum,"[18] which means "iron" in English. Iron accounts for more than 90% of the metals we use, and it is a central element that supports our daily lives. Without iron, we could not make railways, ships, or cars, and it would be difficult to build roads, high-rise buildings, or long bridges. Also, we could not make most of the machines or tools we use today. Therefore, people call iron the "backbone of world industry."[19]

2.3 BEGINNINGS OF THE IRON AGE – BLOOMERY AND ANCIENT STEELMAKING

During the Bronze Age, blacksmiths realized that bronze was inferior in strength and malleability to meteoritic iron, the most precious treasure of the time. They made continuous efforts to make a material similar to meteoritic iron. At that time the monarch of the empire wanted more iron for powerful army weapons and gave a great preference to blacksmiths who could smelt iron. These technologies were passed down primarily through oral communication and apprenticeship experience.[20, 21]

The beginning of the Iron Age overlapped with the Bronze Age.[22, 23] Chalcopyrite, a bronze ore widely used at the time, was a Cu–Fe–S compound, and the slag, a by-product of copper smelting, always contained iron. Sometimes archaeologists consider this reduced iron found in the slag to be an iron artifact from the Iron Age. Iron will corrode and oxidize over time. As a result, iron artifacts preserved for thousands of years have been damaged by corrosion, making it difficult to estimate the exact age of manufacture and use. For this reason, the source of the artifact is presumed based on the results of investigations of other relics that have been preserved in the same area and preserved records. Since the temperature at which iron ore is reduced is much higher than that of copper, stronger winds are needed. Anatolia in Türkiye was a highland region where strong winds frequently blew and there were many iron ore mines. Therefore, the area had very good natural conditions for the opening of an iron civilization.[23, 24]

The bloomery process shown in Figure 2.8 is an iron smelting method that has been used for a long time.[25] This process continued to be used for small-scale production of iron long after the commercialization of the

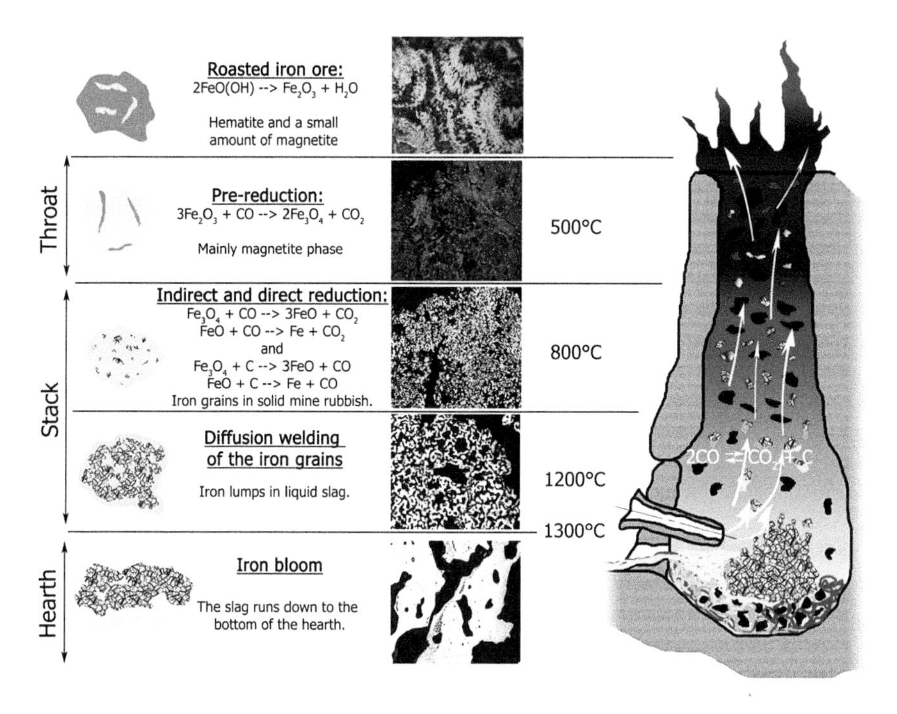

FIGURE 2.8 Iron smelting metallurgy in the bloomery furnace.[25] (Courtesy of http://www.pp.bme.hu/me.)

blast furnace. It involved heating and reducing iron ore with charcoal in a small kiln. The bloomery furnace was typically a kiln with a radius of approximately 1 m and a height of about 1 m. The kiln was made of refractory materials such as clay or stone, and the inner wall was coated with mud. Air was supplied to the furnace using a hand-operated bellow. Good ventilation was crucial when constructing the bloomery furnace. At that time, the primary concern of blacksmiths was controlling the air supply to maintain combustion for an extended period.

The choice of wood for charcoal was also a concern of blacksmiths. In the case of the Middle East where the early Iron Age began, hardwood trees such as acacia and pistachio were preferred as the primary sources of raw material for charcoal.[26] The iron ore was surrounded by charcoal in a bloomery furnace with a chimney at the top to emit gas and holes at the bottom to supply air. Charcoal was first charged into the bloomery furnace, and then iron ore and charcoal were charged in the layer. After the ignition of charcoal, the air was injected through a bellows.

When carbon is burned, it reacts with air to produce CO and CO_2 gas. Gaseous CO is very unstable and reacts with surrounding oxygen or oxygen compounds to become more stable CO_2. This is the main reaction in which carbon reduces metal oxides into metals. When animals inhale CO gas, CO reacts with iron, the main component of hemoglobin in the blood, and deprives them of oxygen, resulting in oxygen deficiency symptoms and fatal results.

When the internal temperature of the bloomery furnace exceeds about 800°C, carbon in the charcoal reacts with the air to produce carbon monoxide as shown in Equation (2.2) below. Thermodynamic data here were referenced from the FactSage FactPS database.[27]

$$2C + O_2 = 2CO, Q = 54.35 \text{ kcal @ } 1{,}273K \quad (2.2)$$

When carbon monoxide passes through the ore bed, the iron ore is reduced to produce iron.

$$3CO + Fe_2O_3 = 2Fe + 3CO_2, Q = 8.37 \text{ kcal @ } 1{,}273K \quad (2.3)$$

The composition of iron reduced in this way is similar to that of pure or low-carbon iron. However, gases such as CO and CO_2 cannot completely escape inside the reduced iron in the bloomery process, which results in the formation of semi-molten iron lumps with pores. This iron is known as bloomery iron. The method of producing reduced iron at about 1,200°C by carbon monoxide is referred to as a low-temperature reduction method or a direct reduction method.[28]

The temperature achieved through charcoal burning in the bloomery furnace was not high enough to melt iron, which has a melting temperature of 1,538°C. In a bloomery furnace, slag with various impurities could be melted. However, the bloomery iron was in a semi-molten state, and the final products were a mixture of iron and slag. The amount of iron product was quite small compared to the amount of ore input.[29] By hammering this sponge iron lump to remove the internal slag and compressing the pores, so-called wrought iron with a carbon content of less than 0.1% can form a bloom. On the cross-section of the wrought iron, the slag that has not yet escaped is distributed in very long stringers of inclusions with iron-silicate composition due to processing such as hammering.[30] It has a vulnerability that can be easily separated. Also, wrought iron with low-carbon content has low strength, hardness, and toughness, which makes it unsuitable for tools or weapons. Therefore, additional steelmaking including chemistry modification was necessary to produce tools or weapons stronger than wrought iron or bronze products.[31]

The old bloomery process has been gradually improved since its initial application. As the design of the smelting furnace was changed from the old bloomery type to BF type, the furnace temperature increased. When the temperature inside the BF exceeded about 1,000°C, carbon reacted with oxygen in the air to generate a lot of heat (compare with Equation (2.2)).

$$C + O_2 = CO_2, Q = 94.45 \text{ kcal @ } 1{,}273\text{K} \tag{2.4}$$

At a temperature higher than 1,300°C this CO_2 reacts with carbon again as shown in the following equation to generate CO gas.

$$CO_2 + C = 2CO, Q = -40.10 \text{ kcal @ } 1{,}273\text{K} \tag{2.5}$$

This reaction is called the Boudouard reaction, named after its discoverer, Octave Leopold Boudouard (1872~1923) of France.[33] The Boudouard reaction is reversible, and carbon monoxide can be decomposed into carbon dioxide and carbon in the furnace. The decomposed carbon carburizes the iron, resulting in a lowering of the melting point as shown in Equation (2.6) below. This carburizing process is called cementation. Carbon monoxide generated by the reversible reaction reacts with oxygen at a temperature as shown in Equation (2.7), generating carbon dioxide and heat to melt iron.

$$C + Fe = Fe[C] \tag{2.6}$$

$$2CO + O_2 = 2CO_2, Q = 134.56 \text{ kcal @ } 1{,}273\text{K} \tag{2.7}$$

The method of producing molten iron in this way is called the indirect or high-temperature reduction process.[33] Molten iron with a carbon content of 4.3% produced in this way lowers its melting point to 1,147°C and is filled in the lower part of the furnace in a molten state, as shown in the Fe–C phase diagram in Figure 1.7.

In the high-temperature reduction method, not only a lot of heat is obtained by the oxidation of carbon and carbon monoxide, but also carbon generated by the decomposition of carbon dioxide can diffuse into molten iron. Achieving high temperatures in the old bloomery process was a very difficult task. It is reported that technology to secure a high temperature has been continuously developed since the 5th century BC in all regions including Rome, but the temperature was not high enough to mass-produce molten pig iron except in China.[30, 31, 34]

The records of how wrought iron was converted to steel during the early Iron Age, beginning around 1200 BC, have rarely been reported yet. However, among the excavated iron artifacts, the surface carbon contents were higher than those in the wrought iron. Therefore, it was assumed that some steelmaking was attempted in one way or another to have a higher carbon content on the surface of the artifact. It is assumed that two methods were used. One is to increase the charcoal content and change the blowing angle while using the same bloomery furnace to raise the temperature further to generate carburizing as shown in Equation (2.6). The steel produced in this way was named "Natural Steel" by metal historians.[31, 32] Another method is to bury the wrought iron in charcoal in a separate furnace and heat it by blowing air to diffuse the carbon or CO gas of the charcoal to the wrought iron. This is to increase the carbon content as in Equation (2.6) or Equation (2.8). This method was widely used in Europe when crucible steelmaking or puddling processes were common.[31–33]

$$\mathrm{Fe} + \mathrm{Fe_XO}\ \text{(partial reduced)} + \mathrm{CO} = (1{+}\mathrm{X})\mathrm{Fe[C]} + \mathrm{CO_2} \qquad (2.8)$$

Although the resulting steel products were not uniform in carbon distribution, they had a high carbon content, so they could be made into tools or weapons with strong steel. This process is called the cementation process.

In the case of ancient China, steel was made by a different process because blast furnaces were used instead of bloomery furnaces. Pig iron from a BF has a carbon content of more than 3%, so the material is so brittle that it could be used as it is for some limited casting purposes, but a process to dilute carbon was required to make tools or weapons. In China, pig iron was melted and a lump of wrought iron was submerged. The surface of the lump could be hardened by diffusion of carbon from the pig iron to the wrought iron. This method spread from China to Korea and Japan. In some cases, this process was repeated more than 100 times to make a famous sword.[35, 36] This process was later named co-fusion.[37] The temperature is raised until the surface of the wrought iron lump melts. And when air is injected, the carbon in the pig iron reacts with oxygen to generate heat. If air is injected while stirring this melting pig iron lump, the decarburization reaction continues and it is converted into steel. This method was named the steel frying process.[35, 38] This principle evolved into Bessemer's converter technology.

These processes of making steel used in ancient times required a lot of manpower, time, and charcoal, while the material of the products made was sometimes uneven. Measurements of the most important variables of this process, such as temperature, wind strength, wind direction, and heating time, also depended on the experience of blacksmiths. In addition, the development of these processes was very slow because no documents specifying these processes were left behind.

REFERENCES

1. Mark R. Leach, "The INTERNET database of periodic tables", https://www.meta-synthesis.com/webbook/35_pt/pt_database.php?PT_id=593
2. "Big history", last edited March 20, 2024, https://en.wikipedia.org/wiki/Big_History
3. "Big bang", last edited March 11, 2024, https://en.wikipedia.org/wiki/Big_Bang
4. S. H. Kim, "Big History of Seohyung Kim -Chronology of Fe", Eastern Asia, 2017 (in Korean)
5. S. W. Kim and J. H. Kim, "Studying physics beginning with big history-start the Big Bang journey", Childeren are Nature, 2018, pp. 12–15 (in Korean)
6. "Planet Earth", WiseUp Library, *Mobius*, Vol. 2, 2002, pp. 5–10 (in Korean)
7. Jennifer A. Johnson, Brian D. Fields, and Todd A. Thompson, "The origin of the elements: A century of progress", *Phil. Trans. R. Soc. A*, Vol. 378, p. 20190301, http://dx.doi.org/10.1098/rsta.2019.0301
8. "Composition of the universe – element abundance", *Science Notes*, https://sciencenotes.org/composition-of-the-universe-element-abundance/
9. Gamow, G., "The origin of elements and the separation of galaxies", *Phys. Rev.*, Vol. 74, No. 4, 1948, pp. 505–506. Bibcode:1948PhRv.74.505G, doi:10.1103/PhysRev.74.505.2
10. E. Margaret Burbidge, G. R. Burbidge, William A. Fowler, and F. Hoyle, "Synthesis of the Elements in Stars". *Rev. Mod. Phys*, vol. 29, 547, Published 1 October 1957
11. "How Many Elements Can Be Found Naturally?", *ThoughtCo.*, https://www.thoughtco.com/how-many-elements-found-naturally-606636
12. Tereza Pultarova, "'Cow' supernova is brightest ever seen in X-ray observations", January 13, 2022, https://www.space.com/bright-cow-supernova-shines-in-xray
13. M.M. Woolfson, "Solar System – its origin and evolution", *Astron. Soc.*, Vol. 34, 1993, pp. 1–20

14. "Abundance of the chemical elements", last edited March 17, 2024, https://en.wikipedia.org/wiki/Abundance_of_the_chemical_elements#Earth
15. Hobart M. King, Ph.D., RPG, "Iron ore", https://geology.com/rocks/iron-ore.shtml
16. "Iron, earth sciences for Australia's future", https://www.ga.gov.au/education/classroom-resources/minerals-energy/australian-mineral-facts/iron
17. U.S. Geological Survey, "Mineral commodity summaries", 2023, https://www.usgs.gov/centers/national-minerals-information-center/mineral-commodity-summaries
18. "Iron", last edited March 27, 2024, https://en.wikipedia.org/wiki/Iron
19. "INFOGRAPHIC: Steel: The backbone of world industry", https://www.mining.com/web/infographic-steel-the-backbone-of-world-industry/
20. "The history of blacksmithing", https://www.forgemag.com/articles/84597-the-history-of-blacksmithing
21. "The life of an iron age blacksmith", https://workingtheflame.com/iron-age-blacksmith-guide/
22. "First smelting of iron", https://ethw.org/First_Smelting_of_Iron
23. T. A. Wertime, "The beginning of metallurgy: A new look", *Science*, Vol. 182, No. 4115, November 30, 1973, pp. 875–887, https://www.science.org/doi/10.1126/science.182.4115.875
24. Ünsal Yalçın and Hadi Özbal, "History of mining and metallurgy in Anatolia", https://www.tf.uni-kiel.de/matwis/amat/iss/kap_a/articles/anatolian_metallurgy_review_yalcin_oezbal.pdf
25. Adam Thiele, "Smelting experiments in the early medieval fajszi-type bloomery and the metallurgy of iron bloom", *Mech. Eng.*, Vol. 54, No. 2, 2010, pp. 99–104, doi:10.3311/pp.me.2010-2.07, http://www.pp.bme.hu/me
26. Lee Horne, "The role of charcoal and charcoal production in ancient metallurgy", https://www.penn.museum/documents/publications/expedition/PDFs/25-1/Fuel.pdf
27. FactSage FactPS database, NIST JANAF Thermodynamic data
28. "Direct reduced iron", last edited February 3, 2024, https://en.wikipedia.org/wiki/Direct_reduced_iron
29. Bernard Graves "Iron working – The bloomery furnace & smelting process", *Description of Iron Smelting using the Bloomary Process*, Publication from Pyrites, Spring 2005, pp. 1–45.
30. R. F. Tylecote, "A History of Metallurgy", 2nd ed., The Institute of Materials, 1992, https://www.academia.edu/40301278/HISTORY_OF_METALLURGY_2nd_Edition
31. K. C. Barraclough, "The development of the early steelmaking processes an essay in the history of technology", Thesis submitted to the University of Sheffield for the Degree of Doctor of Philosophy, 1868, https://etheses.whiterose.ac.uk/14433/1/237901_vol.1.pdf
32. "Boudouard reaction", last edited April 9, 2022, https://en.wikipedia.org/wiki/Boudouard_reaction

33. H. R. Schubert, “History of the British Iron and Steel Industry from 450 B.C. to A.D. 1775”, London and Beccles: William Clowes and Sons, Limited, 1955, https://archive.org/details/in.ernet.dli.2015.104051
34. M. H. C. V Landrin Jr., “A treatise on steel comprising its theory, metallurgy, properties, practical working, and use”, Translated from The French, https://www.loc.gov/item/06018711/
35. “Chiljido” (in Korean), last edited April 20, 2024, https://ko.wikipedia.org/wiki/%EC%B9%A0%EC%A7%80%EB%8F%84
36. Tang Di, “Fried steel and hundred steelmaking: The steelmaking technology of the Eastern Han Dynasty was so great that it was 1800 years ahead of the West”, *LaiTimes*, February 26, 2022, https://www.laitimes.com/en/article/37jvo_3o9bu.html
37. S. Qiao and W. Qian, “Replication experiments and microstructural evolution of the ancient co-fusion steelmaking process”, *Metals*, Vol. 10, 2020, p. 1261, doi:10.3390/met10091261
38. “How did ancient craftsmen make iron and steel?”, September 7, 2023, https://inf.news/en/culture/4bf791e88c5e1a6bced8eb55949de0bb.html

CHAPTER 3

Manufacturing Technology of Steel

IRON ORE EXISTS IN the oxide, hydroxide, and sulfide compounds of iron. The ore exists typically with other types of metal compounds and gangue. Since gangue hinders the reduction reaction of iron ore or consumes a large amount of energy, steps are taken to increase the purity of iron oxides through physical and chemical beneficiation.

Before the Industrial Revolution when there was not much demand for steel, iron ore was mined from easily accessible open pit mines or iron sand near rivers. As the Industrial Revolution spread in earnest, the demand for iron ore soared, and iron ore with low iron content buried underground had to be mined. The blast furnace was advantageous for mass production compared to the low-capacity bloomery furnace, and its operation was greatly affected by the iron content of the beneficiated ore.[1–3] For this reason, the use of high-purity iron ore was increased for productivity. This resulted in an increased amount of tailings after beneficiation.

The reduction reaction of the beneficiated ore was performed in a bloomery furnace by gases generated while burning charcoal. At about 800°C and above, the charcoal reacts with oxygen in the ore and generates CO gas (Equation (2.2)), producing low carbon iron reduced in the solid state (Equation (2.3)), which was called wrought iron until recently. If there is a sufficient supply of oxygen, charcoal, and oxygen react well to produce CO_2 (Equation (2.4)) with a generation of a lot of reaction heat generated and a high temperature can be secured. When the heated iron ore passes

DOI: 10.1201/9781003419259-3

through the high-temperature region above 1300°C, the Boudouard reaction occurs, and the CO_2 generated turns into CO (Equation (2.5)). This CO later reacts with iron ore generating CO_2 and, at the same time, carburizing the reduced iron (Equation (2.6)). This carburized iron is called pig iron and has a low melting point. When the carbon content is increased to 4.3% by carburizing, the melting point is 1,147°C, which makes it easier for the iron to melt in the reactor and activates the refining reaction.

Today, we refer to the product from a blast furnace as pig iron or cast iron, and the low carbon product from a BOF (Basic Oxygen Furnace) or an electric arc furnace as steel. The word "steel" is known to have been in use since around c. 1200.(4) However, it is not known exactly when the definition of the term "steel" as we use it today began to be used. In 1722, Frenchman René-Antoine Ferchault de Réaumur (1683~1757) experimentally demonstrated the process of converting wrought iron or pig iron into steel through the diffusion of carbon, known at the time as "phlogiston," in a paper entitled "Memoirs on Steel and Iron." The authors believe that this is the beginning of the establishment of the scientific definition of "Steel."(5–7) As seen in the previous chapter, until the Industrial Revolution, wrought iron could be produced in kilograms over several days. However, modern integrated steel mills are capable of mass production of steel, up to tens of thousands of tons per day. In this chapter, we will look at how this innovation process developed in a relatively short period of 300 years.

3.1 BLAST FURNACE AND COKE

Some steelmaking methods such as co-fusion or cementation processes have been developed to produce stronger steel than wrought iron for weapons and household tools. However, as shown in Figure 1.7, to obtain steel with a carbon content of less than 2.0% in a molten state, a temperature of 1,400°C or higher must be attained, but the technology was not yet mature enough. By the beginning of the 13th century, the blast furnace used in China shown in Figure 3.1 spread to Europe starting with Sweden,(8–13) replacing the conventional bloomery process, and technological innovation was achieved to mass-produce molten iron.

The blast furnace (BF) is characterized by its high height. Present evidence suggests that the Swedes were the first to use blast furnaces for the smelting of iron. The two sites, Laphphyttan and Vinarhyttan, were confirmed to have been in operation between AD 1150 and 1350 as a result of dating analysis using radioactive carbon.(14–16) BF ironmaking gradually advanced westward to Belgium and southwestern France, and by the

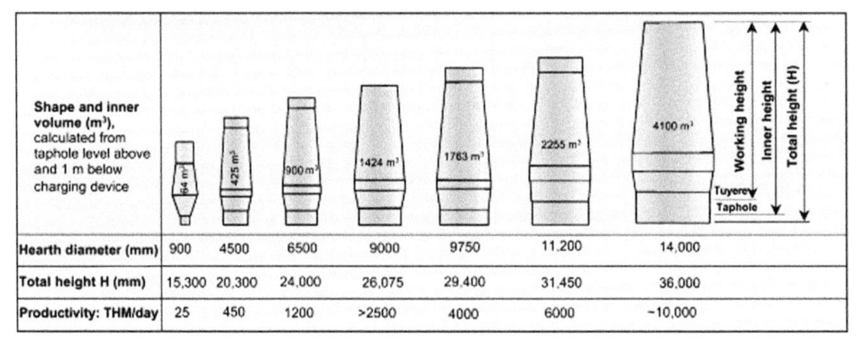

Hearth diameter (mm)	900	4500	6500	9000	9750	11,200	14,000
Total height H (mm)	15,300	20,300	24,000	26,075	29,400	31,450	36,000
Productivity: THM/day	25	450	1200	>2500	4000	6000	~10,000

FIGURE 3.1 Blast furnace in early stage (top left),[13] modern blast furnace (top right),[12] and inner profile (bottom). (Top left, courtesy of http://www.diva-portal.org/smash/get/diva2:1332148/FULLTEXT02, Top right, courtesy of https://commons.wikimedia.org/wiki/File:Alto_Horno,_Puerto_de_Sagunto,_Espa%C3%B1a,_2015-01-04,_DD_91.JPG, Bottom, with permission from Copy Clearance Center, https://www.sciencedirect.com/topics/chemistry/blast-furnace)

end of the 15th century, it had crossed over to the South Sussex coast of England.

The iron industry grew steadily in various regions of England around the 16th and 17th centuries. In particular, ironmaking in the lower reaches of the Severn River in western England, coupled with iron product manufacturing in the upper reaches, created a new large-scale iron industry. At that time, the ironworks with the BF were located in the valley where the water wheel was rotating, and ore processing, refining, and casting factories were located around them.[17] The method of blowing the BF by manpower or livestock was replaced by water power. This was the first improvement for making molten pig iron in a BF from the viewpoint of ensuring high temperature. When a large amount of strong wind is blown into the BF, the combustion of charcoal is accelerated, the temperature is raised and the carburization is activated as shown in Equation (2.5). Referring to Figure 1.7 again, when the carbon content in iron reaches 3 to 4%, the melting point decreases to below 1,200°C. This indicates that the carburized pig iron has a low melting temperature and can exist in a molten state in the BF.

The second improvement to produce molten pig iron involved two measures to optimize the shape of the BF. The first was the reduction of the diameter of the lower part of the furnace tuyere to activate the hotbed and the second was to increase the length from the tuyere to the furnace top, i.e., the furnace shaft. By extending the shaft, the reaction time between the reducing gas and the falling iron ore increased, and the ore was heated by the sensible heat of the generated gas, thereby increasing the thermal efficiency. Consequently, the overall height of the BF was increased to maximize thermal efficiency by accelerating the reaction in the BF using fluid dynamics and thermal engineering. This technology was one of the most groundbreaking inventions in the thousands of years of steel smelting technology.[12]

The production of molten iron opened up a new industry of casting. The foundry industry enabled European empires to arm themselves with cannons and shells of improved performance. Siegerland in the Rhine River basin produced cannons and supplied them to the German government in 1450. Liège in Belgium was also famous for cannons made of cast iron.[17] High-carbon pig iron has low toughness and is very brittle. Reducing the carbon content in molten pig iron improves toughness, making it possible to produce malleable cast iron with good workability. This method is called the indirect process because the molten iron made in the BF is produced by blowing oxygen in a separate furnace, which is different from the direct process used to make wrought iron in the bloomery process. However, during the process of making malleable cast iron, the carbon content decreases, and the melting point increases. At that time, there were no means to secure such a high temperature, so malleable iron was in a semi-molten state.

The iron and steel industry using BFs began in England at the end of the 15th century, later than in other European countries, but it progressed much faster over the next 100 years. BF construction was concentrated in Sussex and Kent, where there were extensive forests. The number of furnaces in Sussex was only two in 1500. But the government's desire to increase military power in a short period greatly motivated the BF construction and, in 1548, the number increased to 53.[9]

In the 16th century, England produced large-caliber cannons with increased range and explosive power using high-quality malleable iron with controlled carbon content. By completely expelling German merchants who had entered the Hanseatic League with the newly manufactured large cannon, Britain was able to protect its commercial rights in northern Europe. In the Age of Exploration, she defeated the Spanish Armada and became the ruler of the sea.

The invention of cannons in the 13th century and firearms in the 14th century fueled a thirst for quality steel. Malleable iron, produced by the indirect method, was poured directly into the molds of gun barrels, and Europe began to mass-produce many weapons. In England, the iron processing industry developed around the BF area of Sussex. Large-scale commerce was created by supplying iron not only to England but also to the rest of the world. The steel-related industry created at this time was as important as the wool industry, which was the most important industry in England at the time, and more than 60,000 people worked in the iron and steel industry.

The amount of wood required to operate one BF in England was equivalent to about 1 million square meters of forest. As iron and steel production nearly depleted timber reserves in the home country, England reduced domestic production and began importing iron smelted from Russia, as well as from Western Europe, Sweden, and the American colonies. Accordingly, the number of BFs in England rose from 58 in 1574 to 85 in 1600 but decreased to 61 in 1717 and 24 in 1790.[9]

Attempts to use coal instead of charcoal as a fuel for iron production began in the 16th century. However, if coal was used as it was, there was a problem in that sulfur contained in coal weakened the mechanical properties of steel products. As a solution to this, Abraham Darby (1678–1717), a cast iron cook, developed a method for producing low-sulfur coke in 1709. Darby's coke was made by putting coal in an oven and roasting it at a high temperature of 1,000~1,300°C for a long time to significantly lower the sulfur content.[18]

One day Darby received an order to make a cylinder from Newcomen (1663–1729), who was developing a steam engine to be used as a pump for coal mines. Darby thought that improving the quality of the iron was essential to making a good cylinder, but at the time, the ironworks could not properly produce suitable iron. Darby decided to venture into the steel industry on his own and chose Coalbrookdale as the site to build an ironworks. Coalbrookdale was adjacent to mountains rich in iron ore and was an area where coal with low sulfur content was easy to mine.

After six months of trials, Darby succeeded in developing a coke-making method in 1709 and was able to mass-produce iron using this coke as fuel.(18, 19) In the developed method, coal was carbonized at high temperatures to make coke. At the same time, iron ore and coke are in contact for a long time, and the coke converts carbon monoxide in the BF to sufficiently reduce the iron ore. This coke-making method developed by Darby continued technological evolution through his descendants, Darby II and Darby III, and saw the establishment of full-scale ironmaking by coke and BF in 1735. Darby III built the world's first Iron Bridge in Figure 3.2, with this cast iron. The Iron Bridge reached 60 m in length over the Severn River in Coalbrookdale, was active until the 1950s, and is still preserved in its original form.

Coke played a sufficient role as both a reducing agent and a heat source for smelting iron ore in the BF. Darby's roasted coal, or coke, retained heat for a longer time than charcoal, which allowed metallurgists to create liquid pig iron ideal for pouring into cannon molds. The BF was built by Darby, as shown in Figure. 3.3, and is now preserved at the Coalbrookdale Iron Museum.

FIGURE 3.2 The first Iron Bridge in the world. (Courtesy of Wikimedia Commons, https://commons.wikimedia.org/wiki/File:Iron_Bridge_east_side_in_February_2019_(cropped).jpg)

FIGURE 3.3 Darby's blast furnace on display at the Coalbrookdale Museum of Iron. (Courtesy of Wikipedia Commons, https://commons.wikimedia.org/wiki/File:Darby_furnace_UK.jpg)

3.2 STEELMAKING AFTER 1700S

The pig iron produced through the ironmaking process has a high carbon content and contains impurities such as phosphorus, sulfur, and silicon, making it brittle. To convert pig iron into tough and high-strength steel, the carbon content must be decreased and harmful impurities must be removed. The methods used since ancient times to make steel include cementation, the co-fusion method, the steel frying method, and crucible steelmaking in India. Later, the crucible steelmaking method and puddling were developed, and the scale of producing high-quality steel gradually increased. Table 3.1 summarizes the evolution of the process of converting iron to steel since ancient times.

The cementation process, as already described in Equations (2.6) and (2.8), is a method in which carbon monoxide or carbon reacts with oxygen in iron and free carbon is carburized to make steel.[20–22] The co-fusion and steel frying steelmaking methods are processes of manufacturing steel by mixing molten cast iron made using the high-temperature reduction method with wrought iron and iron ore, respectively. Since these methods had a disadvantage in that the melting point gradually increased as the content of carbon decreased and may solidify, it was important to maintain a high enough temperature.

TABLE 3.1 Steelmaking Process Evolution Chronology.

Process		Inventor	Year
Small scale	Cementation	Hittite?	BC 1800?
	Co-fusion	Northern and Southern Dynasties	BC 500~600
	Steel frying	Han dynasty	BC 200
	Crucible steelmaking	India	BC 500?
Medium scale	Crucible steelmaking	Benjamin Huntsman	1740~
	Puddling	Henry Cort	1783~
	Induction furnace	Michael Faraday	1831~
Mass production	Bessemer converter	Henry Bessemer	1855~
	Open-hearth furnace	Carl Wilhelm Siemens and Pierre-Émile Martin	1865~
	Thomas converter	Gilchrist Thomas	1877~
	Electric arc furnace	Carl Wilhelm Siemens	1878~
	LD converter	Robert Durrer	1953~

3.2.1 Benjamin Huntsman's Crucible Steelmaking

Benjamin Huntsman (1704–1776), who was working as a clockmaker in Sheffield, England, was always dissatisfied with the performance of springs used in watches. He wondered if he could make better steel than the existing cemented steel for clock springs at that time, and developed the crucible process, shown in Figure 3.4. Experimenting with smelting iron in different ways, Huntsman found a process very similar to the crucible steelmaking method used to make Wootz steel in ancient India.[20, 23]

The crucible steelmaking method used pig iron from a BF or wrought iron from a bloomery furnace. It is a method of making molten steel in the crucible (C) of Figure 3.4, burning charcoal or coke in H, and supplying CO through channel G while heating the crucible, melting and adjusting the carbon content of the iron in a container. Composition control was also possible by charging ferroalloy or steel with different compositions into the crucible. Crucible steel made by this method was superior in terms of purity, uniformity, and quality compared to cemented steel made by the conventional method, but the amount produced at one time was very small, only tens of kilograms, and took a long time to manufacture.

The crucible steelmaking method laid the foundation of modern steelmaking methods. Huntsman's persistence in making an accurate clock led to the development of steelmaking technology. After the commercialization

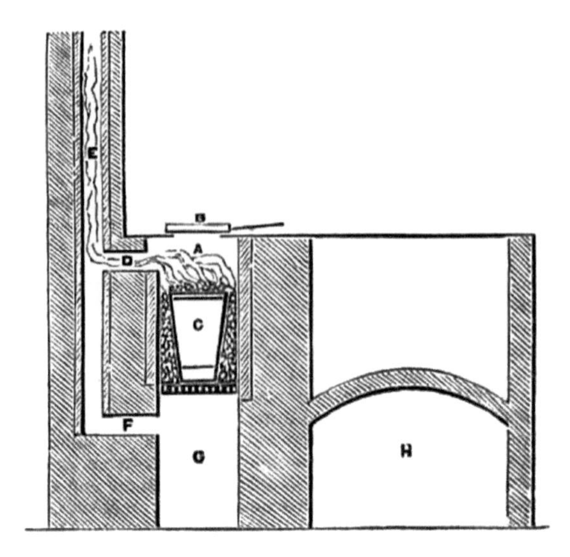

FIGURE 3.4 Schematic of crucible steelmaking developed by B. Huntsman. (Courtesy of https://commons.wikimedia.org/wiki/File:Huntsman_crucible_furnace.png)

of the crucible steelmaking method, Sheffield became a hub for high-grade steel manufacturing after 1770.[20, 24]

3.2.2 Henry Cort's Puddling

Henry Cort (1741–1800) was a ship dealer in London who imported iron from Sweden and Russia to supply to the navy. Cort's job made him aware of the quality problems of British iron compared to foreign iron. In 1775, Cort built an ironworks at Fontlay, Portsmouth Harbour, and in 1784 invented a kind of reverberatory furnace called a puddle, shown in Figure 3.5. He later succeeded in converting pig iron into the steel by the blasting of a forced air.[20, 25]

The puddle method used coal developed by G. T. Cranage and Peter Onions, engineers at the Coalbrookdale Ironworks. The method was based on the reverberatory furnace method. The pig iron inside the furnace was indirectly melted by the reflected heat of the flame, so the sulfur in the coal did not permeate into the iron. The carbon in molten pig iron was removed by oxidation to convert steel. As the iron lost carbon, its melting point rose with some loss of fluidity, and the molten iron should be puddled to activate the reaction. This method was similar to the co-fusion steelmaking method developed in ancient China.

FIGURE 3.5 Schematic of reverberatory furnace for puddling.[25] (Courtesy of https://commons.wikimedia.org/wiki/File:Puddling_furnace.jpg)

The product refined by the puddle method was also called puddled steel. Its quality was excellent due to internal homogeneity. Henry Cort also established a process for manufacturing products by rolling a plate or rod through multi-pass powerful rolling rather than by conventional hammer forging. The rolling mill at that time was driven by a steam engine.

3.2.3 Krupp's Cast Steel History

Krupp is a heavy industry company based in Essen, Germany, with a long history. In 1999, after merging with Thyssen, ThyssenKrupp was born, and its main products include steel, machine parts, elevators, and industrial solutions. Founded in 1810, Krupp succeeded in commercializing industrial machinery by developing casting technology linked to the crucible steelmaking method. Until then, the products were limited in size and shape because blacksmiths had no choice but to forge them on an anvil with hammers. Cast iron had a high degree of freedom in making big things, but it was brittle and weak. At the end of the 18th century, British steel-casting technology monopolized world demand.

Prussian inventor Friedrich Krupp (1787–1826) began research on the British monopoly manufacturing technology for cast steel. He established a research institute and a factory on the banks of the Rhine and developed

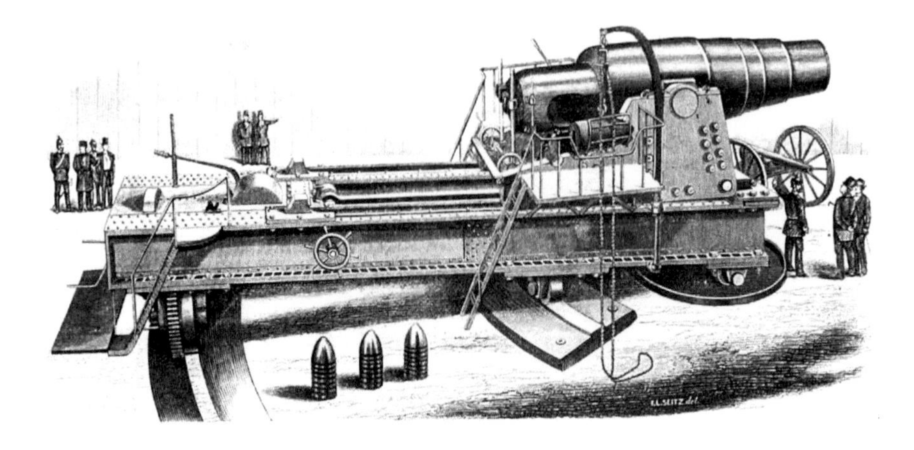

FIGURE 3.6 Cannon made by Krupp in 1870. (Courtesy of https://commons.wikimedia.org/wiki/File:Krupp_50_ton_gun_-_Scientific_American_-_1870.png)

the manufacturing technologies of cast steel for cannons. Based on this achievement, his factory was capable of manufacturing cast steel with power from a wooden waterwheel. Later, he left the factory to his son, Alfred Krupp (1812–1887).

Alfred Krupp manufactured a cannon with an inner diameter of 9 cm from cast steel in 1856, and the results were so impressive that Prussia decided to make military guns from cast steel. Prussia was the first country to make this decision, and Krupp produced excellent cannons one after another. Krupp cannons began to be purchased in large quantities in the 1860s in Russia, Austria, and the Ottoman Empire because of their superior performance compared to conventional cannons. By the 1870s, they were being purchased by countries all over the world. Figure 3.6 shows the Krupp cannon, which was developed not only for ground use but also for naval warships, and the steel used in manufacturing the cannon was cast steel made by the crucible steelmaking method.(26) By the late 1880s arms manufacturing had increased to about 50% of total production, and at that time Krupp employed about 75,000 people. In 1903, a corporate company was formally established, and it became the world's largest steel products company in both name and reality.(27)

3.3 INNOVATION OF STEELMAKING TECHNOLOGY

3.3.1 Open-Hearth Furnace

By the time the Bessemer converter was introduced, German scientist Carl Wilhelm Siemens (1823–1883) and Frenchman Pierre-Émile Martin

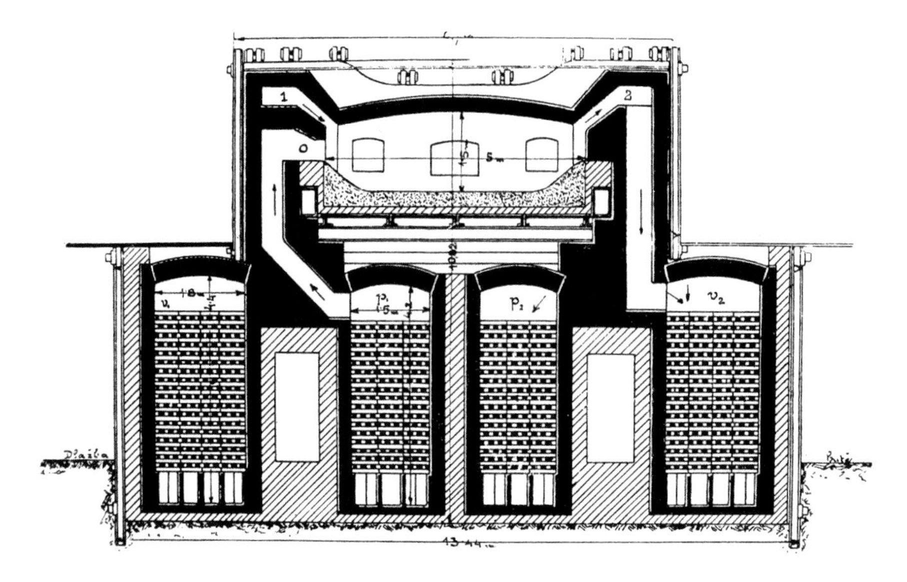

FIGURE 3.7 Siemens open-hearth furnace and regenerators for making steel.[29] (Courtesy of https://commons.wikimedia.org/wiki/File:Siemens-Martin_furnace_%E2%80%93_longitudinal_section,_Otto%27s_Encyclopedia.jpg)

(1824–1915) had developed another efficient method for making molten steel. The existing puddling method required puddling in the latter half of molten iron decarburization due to temperature limitations, and this process was the bottleneck that reduced productivity. Siemens developed a furnace that could omit puddling by raising the temperature by 200 to 300°C higher than the existing reverberatory furnace.

The method they employed to raise the temperature is the heat storage method. As shown in Figure 3.7, this method uses the sensible heat of exhaust gas to reheat and supply combustion air. Siemens succeeded in raising the temperature by redesigning the structure of the heat storage chamber. The high-temperature gas generated inside the melting furnace was discharged to the outside through a pipe, mixed with air, and then returned to the heat storage chamber. In the heat storage chamber, these tubes were densely installed in a checker shape to increase heating efficiency.

Siemens initially encountered challenges as the refractory used could not withstand the high-temperature atmosphere. Siemens met Martin of France around this time, and they collaborated to renovate the heat storage chamber and successfully process molten steel. The development of this technology took nearly 20 years from the conception of Siemens' regenerative furnace to the production of molten steel. For a long time, Martin, along with other experts in France, even organized a national committee to develop their steelmaking technology. By combining his rich experience and knowledge with those of Siemens, the open-hearth furnace was finally successful.[28]

The additional supply of recycled heat from the melting furnace allowed the iron to remain liquid longer than in the Bessemer converter and could even melt steel scrap. This technology not only reheats molten steel by obtaining a high temperature but also maximizes the scope of molten steel refining by activating chemical reactions such as the decarburization of molten steel using oxygen from iron ore. Along with the Bessemer converter, the open-hearth furnace became two major methods for modern molten steelmaking, spreading widely from France to Germany, Russia, and the United States.

3.3.2 Bessemer Converter

In the 19th century, European countries, led by England, began to adopt coke ovens, puddling steelmaking, and rolling methods. England further accelerated the expansion of steel processing facilities through the development of railways in Europe. In 1828, James Beaumont Neilson (1792–1865) developed a method of using hot air instead of cold air,[29] and James Nasmyth (1808–1890) invented a steam hammer.[30] However, the puddle process was not suitable for large-scale steel production due to the difficulties in its mass production, resulting in a mismatch in terms of process integration. The puddling process required a long time to heat up using the indirect heating and reverberatory furnace method, and the high-temperature manual puddling initially had limitations in improving productivity.

Henry Bessemer (1813–1898), an English engineer, was a talented inventor who proposed various innovations such as movable letterpress, sugarcane mill, brass corrosion paint, converter, and strip caster. Among these, his invention of the converter is considered the most innovative, especially for mass production. During the Crimean War (1853–1856), Bessemer built a new long-barreled cannon and supplied it to the French army. Bessemer was concerned about the phenomenon of gun barrels frequently breaking due to the shock at the time of firing. Bessemer believed that producing superior steel was the key to this problem, but the crucible steelmaking method of the time was too expensive to produce large items such as cannons. Thus, Bessemer researched ways to produce high-grade steel cheaply and in large quantities.

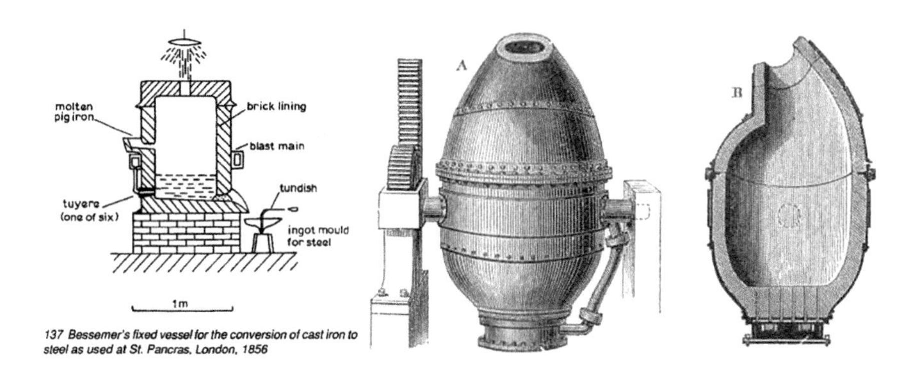

FIGURE 3.8 Bessemer steelmaking process.[32] (Left, courtesy of R. F. Tylecote, *A History of Metallurgy*, 2nd ed., The Institute of Materials, 1992, London, p. 166, Right, courtesy of https://ko.wikipedia.org/wiki/%EB%B2%A0%EC%84%9C%EB%A8%B8_%EB%B2%95#/media/%ED%8C%8C%EC%9D%BC:Bessemer_converter.jpg)

In 1856, Bessemer poured molten iron into a furnace called a converter, as shown in Figure 3.8, and blew air through holes in the bottom of the converter. Then, about 10 minutes later, fireballs and sparks suddenly erupted from the top of the vessel in a 9 m high flame. When the chaos subsided, the material left in the converter vessel was steel, which had a very low carbon content. This technology called the Bessemer steelmaking method, used the heat generated in the oxidation process when the air was blown into the molten iron and carbon was burned in the molten iron through an oxidation reaction.[31]

The great advantage of the Bessemer steelmaking method was a fast processing time, which was only 1/10th of that of the existing technology. While the existing steelmaking method was limited to processing units of 200 kg, the Bessemer converter could work up to 20 tons at a time. In addition, rails made of Bessemer steel had a lifespan of 20 times longer than conventional wrought iron rails. This soon led to a dramatic increase in steel production and was the most cost-effective. Steel became a dominant construction material solely because of this invention. In England, the cost of steel dropped from £40 to £6–7 GBP per long ton.[32]

The puddling method had no choice but to produce high-carbon steel with a low melting point, but the Bessemer converter could easily produce low-carbon steel because the reaction between oxygen and carbon in the molten iron generated heat and a high refining temperature. In addition, since the capacity of the Bessemer converter could be matched to the size of the BF, the basis for the birth of a modern integrated steelworks system that balances production between unit processes was established.

3.3.3 Thomas Converter

Shortly after the introduction of the Bessemer steelmaking process, British steelworks encountered a problem. Most iron ore produced in Britain was high in phosphorus, and this could not be reliably removed by the Bessemer steelmaking method, resulting in steel with poor toughness. However, phosphorus did not cause serious problems in open-hearth furnaces or Puddles. This is because the operating temperature of the Bessemer converter is 200 to 300°C higher than that of conventional steelmaking furnaces, up to around 1,600°C, and P_2O_5 dissolved in slag is rephosphorized into molten steel. Meanwhile, S. G. Thomas (1850–1885), a 25-year-old British court clerk and chemist, found a solution to this problem. Thomas proposed the use of stable basic refractories that would not decompose P_2O_5 at high temperatures but, instead, redistribute this P_2O_5 into slag for discharging.

P_2O_5 is well soluble in CaO, which is classified as a basic oxide. Here, the term “basic oxide” refers to an oxide formed by the combination of oxygen and a metal belonging to groups 1 and 2 of the periodic table, such as Mg and Ca, while the term “acidic oxide” is mainly SiO_2. Initially, the acidic oxide was used as a refractory in the Bessemer converter. However, this acidic refractory reacts with basic slag containing a lot of CaO to form a CaO–SiO_2 compound with a low melting point, leading to a rapid shortening of the converter refractory’s lifespan. To address this issue, Thomas proposed the idea of using basic dolomite as a refractory and using basic slag-containing CaO. This significantly reduced the phosphorus content in the steel.(33)

The use of basic refractories had the additional positive effect of reducing the amount of sulfur entering the steel. For this reason, the basic oxygen steelmaking method became the mainstream of steelmaking, and it was also applied to open-hearth furnace steelmaking, enabling the treatment of iron ore containing phosphorus. This also resulted in the conversion of large-scale mines into resources that were previously unavailable due to the high phosphorus content. A typical example is Minette ore, which has large deposits in the border areas of Germany, France, and Luxembourg. Minette ore contains a lot of phosphorus and little silicon, which makes it an ore suitable for applying the Thomas converter.

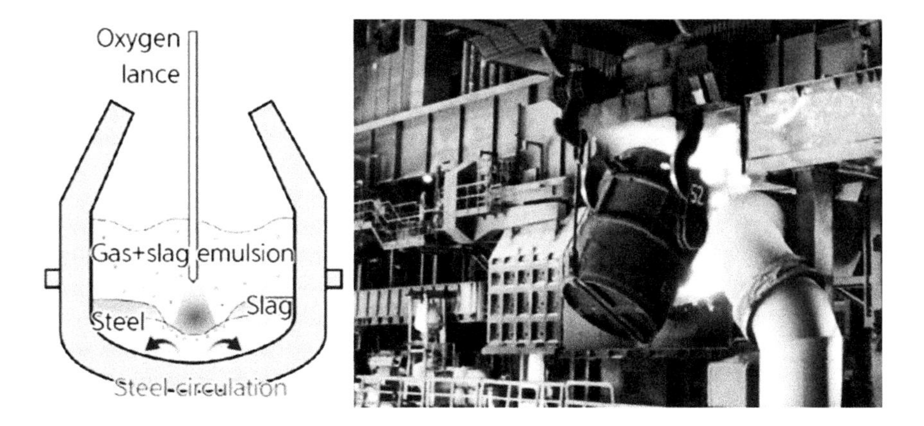

FIGURE 3.9 LD converter process diagram and actual pouring of molten iron to the converter.

3.3.4 Innovative Steelmaking: LD Converter

In the 20th century, a Swiss engineer named Robert Durrer discovered a better way to make steel. Durrer taught metallurgy in Nazi Germany, but after World War II he returned to Switzerland to experiment with the steelmaking process. He refined the process by blowing pure oxygen into the converter instead of air. Upon analyzing the results, he found that carbon could be more effectively removed from molten iron. Molten iron was charged into this converter used in the experiment shown in Figure 3.9, and high-pressure, high-velocity oxygen was blown through a water-cooled lance directly above the molten iron. The converter was commercialized in 1953 at two plants in Linz and Donawitz, Austria, based on Durrer's research in 1946. This process is called the LD converter after the initials of the two plants.(34)

The LD converter steelmaking method uses high-purity oxygen, resulting in lower nitrogen content in the steel compared to open-hearth steel. Additionally, the contact of oxygen with the molten surface generates high temperatures of 2,000~3,000°C, allowing for simultaneous dephosphorization and decarburization while rapidly melting slag-containing lime. Once the carbon content reaches the target and the oxygen blowing is stopped, it produces steel with consistently low phosphorus content. When the LD converter became commercialized, it replaced most of the existing open-hearth furnaces due to its six-fold higher productivity and 30% lower construction cost than the open-hearth furnace.(35) However, switching to converters in Russia and the United States took a long time as many young open-hearth furnaces were still running.

To convert phosphorus-rich molten iron into steel, a method was developed using basic refractories that had been applied to Thomas converters and were then extended to LD converters. Products made in the LD converter are suitable for making thin sheets since they contain fewer impurities that would impair the workability of steel. The heat capacity of the furnace ranges up to 350 tons, making it suitable for mass production, and it has become the mainstream of modern steel refining furnaces.

3.3.5 Electric Arc Furnace and Induction Furnace

The electric furnace process produces steel by directly melting steel scrap using heat generated from electricity. There are two types of electric furnace methods; an electric arc furnace (EAF) and an induction furnace. The EAF melts scrap by arc heat generated between the scrap and the electrode. The IF heats scraps with resistance heat induced by a coil wrapped around a crucible.

The high-frequency induction furnace in Figure 3.10 was invented by Edwin F. Northrop in 1916[36] and is mainly used in industries with a capacity of fewer than 50 tons. This method is used for manufacturing high-alloy special steel such as heat-resistant steel and high-speed tool steel or castings. Induction furnaces are classified into high and low

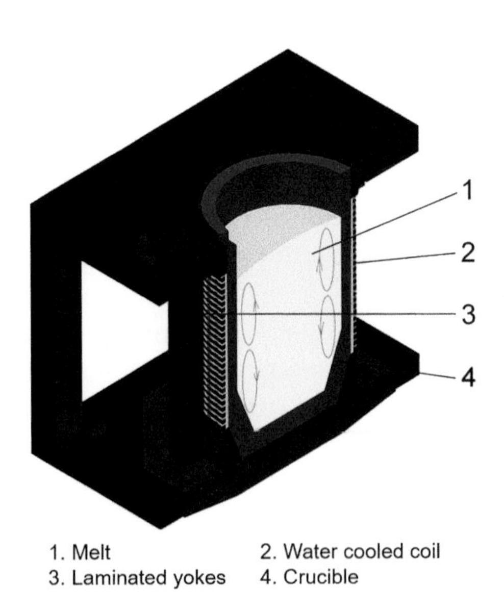

FIGURE 3.10 Induction furnace. (Courtesy of Wikimedia Commons, https://commons.wikimedia.org/wiki/File:Induktionstiegelofen_Schnitt.png)

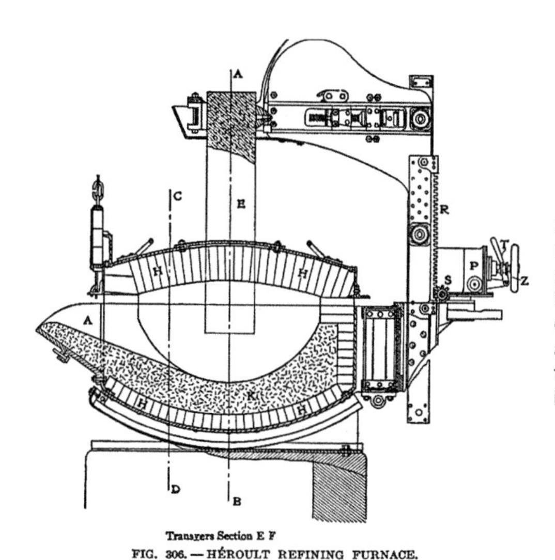

E is an electrode (only one shown), raised and lowered by the rack and pinion drive **R** and **S**. The interior is lined with refractory brick **H**, and **K** denotes the bottom lining. A door at **A** allows access to the interior. The furnace shell rests on rockers to allow it to be tilted for tapping.

FIGURE 3.11 Electric arc furnace. (Courtesy of https://commons.wikimedia.org/wiki/File:Heroult_refining_furnace_Transversal_view_Stoughton.PNG)

frequency, based on the applied current. Due to the long melting time and the indirect heating with the electromagnetic force, the energy efficiency of induction furnace is lower than that of an EAF. Because of this, the EAF as shown in Figure 3.11 is now the primary equipment used in large-volume steel production.

Currently, around 30% of global steel production is produced by electric furnaces, while the rest is occupied by the LD converter. However, these figures vary widely from country to country. The US is the highest at around 60%, followed by the EU at 42%, Japan at 30%, Korea at 38%, and China at around 10%.(37) This difference is mainly due to variations in electricity rates, as enormous power is required to melt scrap in an EAF. The EAF was invented in 1878 by Siemens, who also inspired the production of the open-hearth furnace. In 1899, Frenchman Paul Héroult completed the Héroult-type EAF, which is still widely used today along with its improved form.(38)

In recent years, the importance of EAF has been increasing. For some time, there was a perception that steel produced in EAF could not be used as a high-quality steel sheet for automobiles or high-strength steel for structural purposes. However, this perception is changing with the development of refining technology. Problems associated with tramp elements such as arsenic, tin, and copper in scrap can also be solved by properly mixing and using DRI (direct reduced iron) or molten iron.

Moreover, ultra-low carbon or clean steel can also be manufactured by adding a secondary refining process after EAF and developing appropriate technologies.

In terms of global warming, the EAF route is preferred because the amount of CO_2 emitted is about 25% of that emitted from the BF route. This makes it likely that the production ratio of EAF will increase in the future. Additionally, the amount of scrap recycled from previously produced steel products is continuously increasing and the supply is becoming more stable, making the EAF route even more advantageous. In the case of the US, which has a high proportion of mini-mills, the conditions for using the EAF route are improving. It is also beneficial to secure direct reduced iron (DRI) as is an alternative iron source utilizing cheap shale gas.[39] By expanding the use of EAF, steel production through the BF route can be reduced, which eliminates the need to build a coke and a sinter plant, thereby reducing CO_2 emissions. Moreover, the EAF route is also environmentally friendly as it produces less fine dust, making it more attractive for the steel industry in the future.

3.4 INTEGRATION OF STEEL MANUFACTURING PROCESSES

Looking back over the past 50 years, most of the innovative changes in steel processing have been directed toward improving product quality and reducing manufacturing costs. An important core concept applied for this purpose has been process reduction and integration. Continuous casting, hot direct rolling, continuous annealing, and continuous hot dip galvanizing are typical examples.

3.4.1 Continuous Casting

As the share of continuous casting (CC) in the steel industry continues to grow, engineers have been pleasantly surprised by its potential for cost-saving through yield improvement, investment saving, energy saving, and process integration. The success of CC has sparked research and development activities aimed at integrating the casting, rolling, annealing, and coating processes.

By the 1960s, steelmaking had greatly increased its productivity with the rapid development of the LD converter technology. However, liquid steel produced in steelmaking was poured into a mold arranged in several rows, as shown in Figure 3.12, and cast. This ingot casting (IC), which has been used for more than 30 centuries, was still the mainstream casting method at that time. In the IC as shown in Figure 3.13, a mold must be

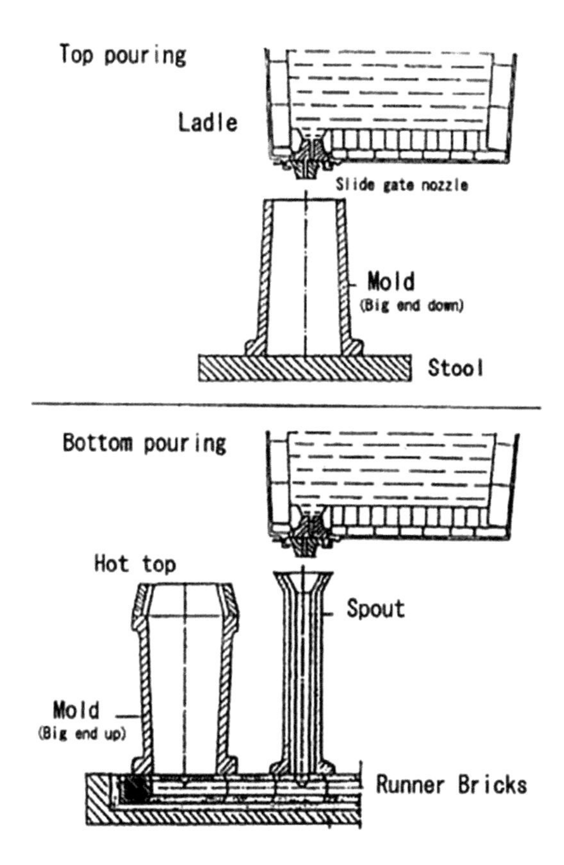

FIGURE 3.12 Top and bottom pouring for conventional ingot casting. (With permission from Steel Data, https://www.steeldata.info/inclusions/demo/help/ingot.html)

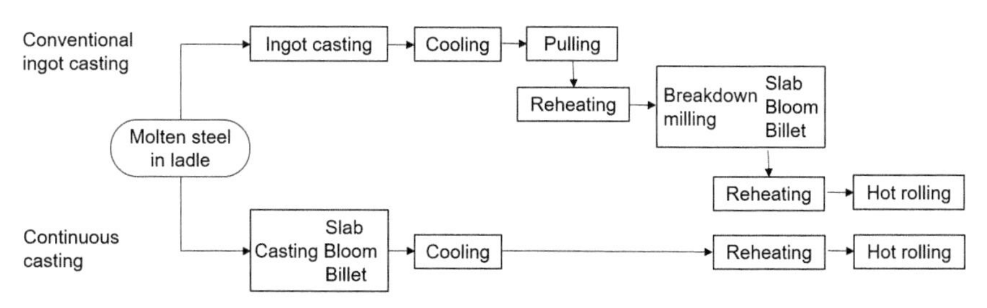

FIGURE 3.13 Process comparison of ingot and continuous casting.

preheated before pouring liquid steel and cooled after casting and pulling out the steel ingot from the mold. This process was discontinuous and consumed a significant amount of time and energy. In addition, the yield was low due to the pipe formed at the top of the ingot and segregation, causing quality deviation in the longitudinal direction.

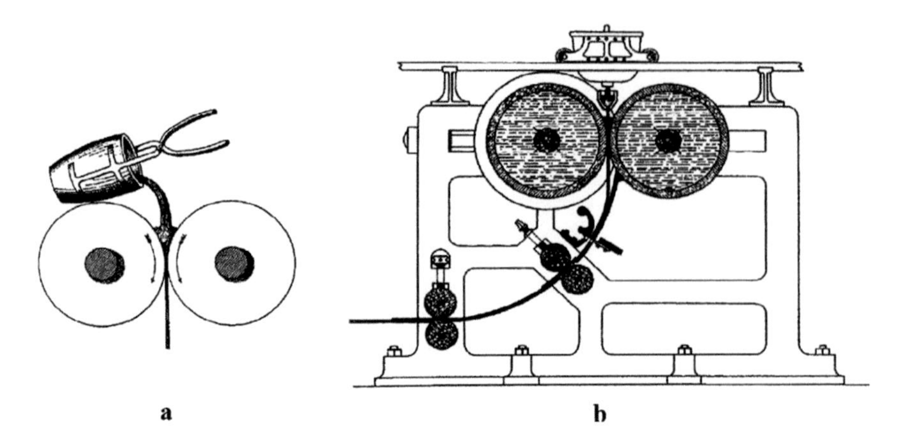

FIGURE 3.14 (a) Bessemer's twin-roll caster design was conceived in 1846 and patented in 1857 and (b) a twin-roll caster for steel, was patented in 1865.[40, 41]

CC has long attempted to solve these issues with IC. In 1850, Bessemer proposed the CC method using twin roll as shown in Figure 3.14. However, Bessemer's twin roll casting (TRC) was discontinued due to difficulties such as the confinement of molten steel on top of the twin cast rolls and the quality of the strip. This concept was revived in the late 1980s with the strip casting process.[40, 41]

In the 1960s, a German named Siegfried Junghans patented the concept of empty mold and mold oscillation and achieved the CC of brass.[42] Since then, CC has developed significantly and has been applied to steel casting. This method is now a key process in the steelmaking process. Figure 3.15 shows the outline of the CC machine where molten steel is continuously poured from the ladle to the tundish and mold, and this liquid steel is initially solidified along the mold. The cast exiting the mold is then solidified as it passes through the cooling zone known as the secondary cooling zone.

Before and after World War II, several countries installed pilot casters to develop the key technologies of CC. The technologies and facilities developed during this time included a mold oscillator, mold powder for lubrication, and shroud and submerged nozzle to prevent reoxidation of liquid steel. Technological development to improve the quality of cast steel continued thereafter, with the development of technologies such as electromagnetic stirring, sliding nozzle technology, nickel coating on mold copper plate, width change during casting, soft reduction technology, high-speed casting, and sequence casting of different steel grade, among

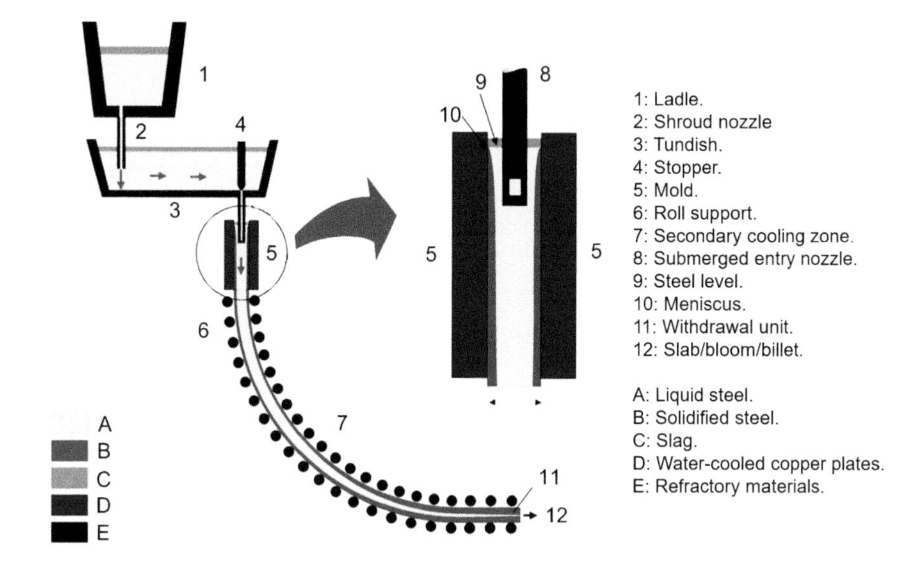

FIGURE 3.15 Concept of continuous casting. (Courtesy of https://commons.wikimedia.org/wiki/File:Lingotamento_Continuo-Continuous_Casting.png)

others. Most of these technologies were developed by steel and engineering companies in Japan and Europe.

If we divide the development of CC technology into stages, the 1950s represent the beginning stage, the 1960s represent the settlement stage, the 1970s represent the development stage, the 1980s represent the maturity stage, and the 1990s represent the golden stage. In the golden stage, high-speed casting was established and most of the existing IC routes were closed due to the successful casting of high-alloy steel, which was previously considered impossible to cast continuously. Key technologies were developed during the beginning stage by experimenting with pilot casters at Mannesmann's Huckingen in Germany, Barrow Steelworks in US Steel (USS), and Sumitomo Metals in Japan.

During the settlement stage, the focus was on commercializing slab casters. In Germany, Dillinger Huette's vertical bending and curved caster at Huckingen Steelworks in Mannesmann was built. Immersion nozzle and mold powder began to be used. The caster of Gary Steelworks in the USS attempted a high-productivity operation of over 2.1 million tons per year.

In the development stage in the 1970s, USS succeeded in CC of pseudo-rimmed steel and introduced tundish sealed casting and Ar gas injection technology were introduced. NSC Oita Steelworks operated all CC processes. A soft reduction of cast strand and high-precision molten steel level control technology in the mold was developed. During this period, as the oil crisis spread across the world and the need for cost-saving became urgent, CC technology with excellent energy-saving effect came into the limelight.

In the mature stage since the 1980s, Japanese steelmakers, which secured the highest competitiveness in the steel industry, synchronized the CC process and the hot rolling process with high-speed casting and hot charged rolling (HCR) technology and clean steel production technology. With the development of various key technologies, it was possible to produce almost all steel products by CC. On the other hand, Nucor of the US has established a basis for securing future competitiveness in the mini-mill steel industry that employs a thin slab caster.

In the golden stage of the 1990s, CC technologies were continuously improved and developed, leading to a high level of technological evolution. The high-speed casting technology of high-alloy steel, as well as cast steel quality improvement technology through advanced flow control of molten steel, have been realized. Figure 3.16 shows the drastic growth of the CC process through the changes in global crude steel production over the 30 years leading up to this period and the rapidly increasing rate of CC.[43, 44]

With the development of high-speed casting technology, CC continues to promote the integration of CC and hot rolling processes. Representative processes being developed accordingly are thin slab casting (TSC) and

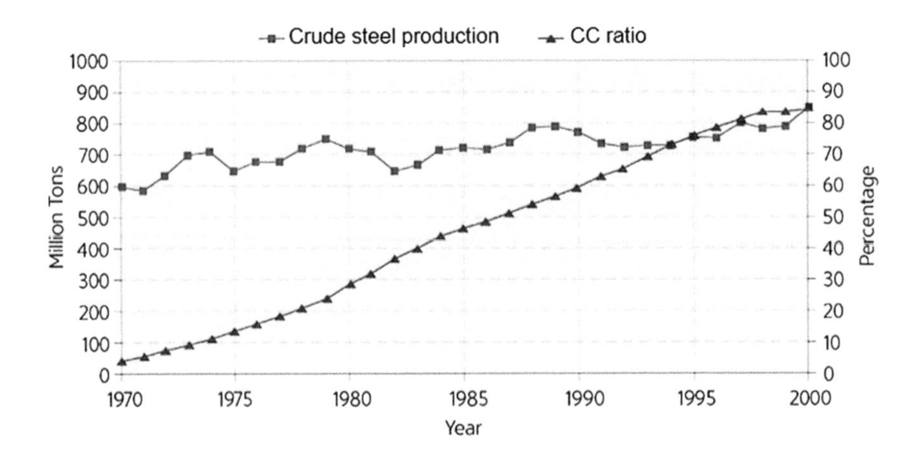

FIGURE 3.16 Increased CC ratio from 1970–2000.

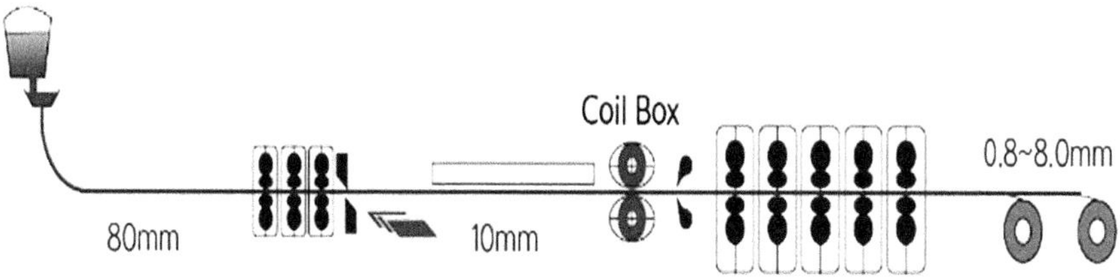

FIGURE 3.17 Concept of continuous casting and direct rolling.

strip casting (SC) technologies.[45] In the TSC process, the thickness of the cast slab is 50 to 100 mm and the casting speed is 3.0 to 8.0 m/min, which is mainly used by EAF steelmakers for producing steel sheets.

More advanced technology has recently been proposed where CC and the hot strip rolling were directly connected as shown in Figure 3.17. This technology was applied not only to the EAF steelmaking process but also to the LD converter steelmaking process. This process was a technology that fundamentally eliminated the re-heating process of cast steel before rolling and will be described in detail in Chapter 8.

In the SC process, which did not require a hot rolling process, as shown in Figure 3.18, the thickness of the cast strip was 2~3 mm and the casting speed was 30~100m/min.[41] This technology has been tried on a commercial scale for stainless steel products in Japan and Korea and for carbon steel in Australia and the US. Commercialization is underway in US Nucor[46] and China; details will be mentioned again in Chapter 8. Since this process has great energy-saving potential, attention is being paid to its future commercialization.

3.4.2 Tandem Rolling

Rolling is the process of plastically deforming metal by passing it between two or more rolls. It is the most widely used forming process and provides

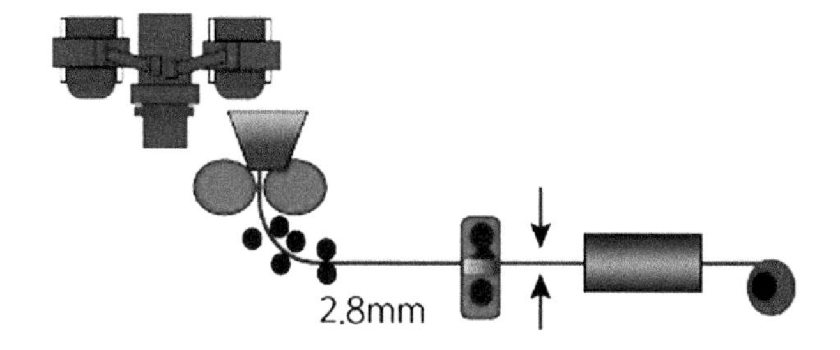

FIGURE 3.18 Typical process arrangement of strip casting and rolling.

FIGURE 3.19 Sketch of a Rolling Mill from Leonardo da Vinci, 1485. (Courtesy of https://www.metalworkingworldmagazine.com/a-short-sheet-metal-history/fig_2-sheet-metal-history/)

high production and close control of the final product. This process shapes metal into a thin, long layer by passing it through a gap of two rolls rotating in opposite directions (clockwise and counterclockwise).

The practice of hot and cold metalworking between two or more rolls has existed since long before the Industrial Revolution. The first known design of a rolling mill dates back to Leonardo da Vinci who, in one of his drawings dated 1485 (see Figure 3.19), describes for the first time the possibility of "making a material pass" between two cylindrical rollers with parallel axes to modify its thickness. The first rolls were small and hand-driven, used to flatten gold and silver in the manufacture of jewelry and art. By the 1600s rolling machines were known to be in operation and steel was just being introduced as a metal capable of being rolled.

The first mention of a tandem rolling mill is Richard Ford's 1766 English patent for the hot rolling of wire. In 1798, he patented the hot rolling of plates and sheets using a tandem mill. The main advantages of a tandem mill are that only a single pass is required, saving time and increasing production. Additionally, greater tension is possible between the stands, allowing for an increased reduction in the stands for the same roll force. One disadvantage of a tandem mill is the high capital cost compared to

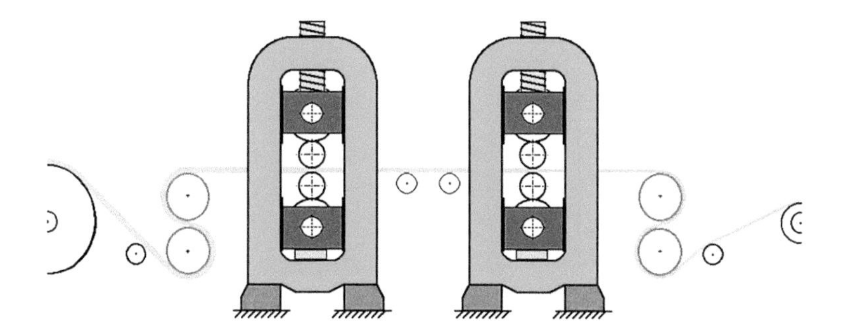

FIGURE 3.20 Two strands tandem rolling mills. (Courtesy of Wikimedia Commons, https://commons.wikimedia.org/wiki/File:Sketch_showing_payoff_reel_(un-coiler),_entry_bridle,_2_stands,_exit_bridle_and_coiler_(tension_reel).jpg)

that of a single stand reversing mill. Due to these characteristics, the tandem rolling shown in Figure 3.20 is applied to hot rolling of sheet and plate, wire rod, and cold-rolling processes regardless of product size in most integrated steel mills currently in operation.[47]

3.4.3 Hot Bar Joining for Endless Hot Rolling

Minimizing the quality deterioration parts can be achieved by joining the tail part of the preceding hot bar roughly rolled in hot rolling, to the head part of the following bar. This also extends the life of the rolls as the bar head deformed during rough rolling does not collide with the rolls. Joining technology has been attempted since the mid-1990s by Japan's NSC and JF, and Korea's POSCO.

Japan's NSC and JFE attempted to join the preceding bar tail and the head of the following bar with a high-frequency induction welder. However, it took a lot of time to pretreat and weld both ends of the bar. Due to the oxidation of the joint, the originally intended advantage could not be obtained and technological development was stopped. Instead of high-frequency induction welding, POSCO tried a shear deformation bonding method, which takes only about 1 second to join. With this technology, most of the problems such as oxidation of the joint that the two companies faced earlier were solved. In 2006, the hot bar joining technology was successfully commercialized.[48] Figure 3.21 is a conceptual diagram of the hot bar joining technology applied at POSCO Pohang Works. The effect is maximized by minimizing the defects during hot rolling and the deviation of material properties that occur in the head and tail.[49] It will be more promising if applied to the hot rolling of electrical steel sheets.

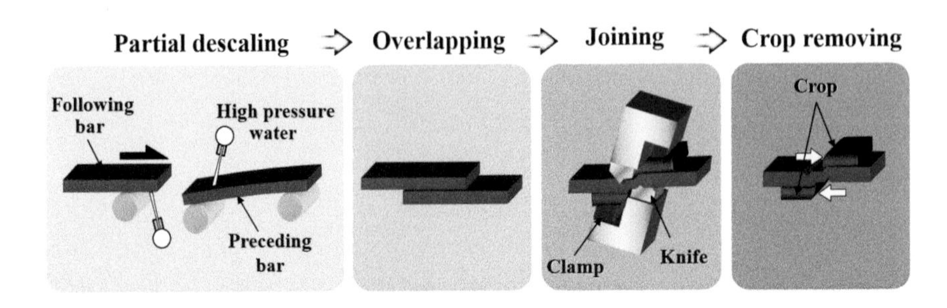

FIGURE 3.21 Schematic diagram of hot bar joining technology commercialized at the hot rolling mill at POSCO.

3.4.4 Continuous Annealing and Hot Dip Galvanizing

When a hot-rolled steel sheet is subjected to a cold-rolling process, it is transformed into a steel sheet with high hardness and low workability. To eliminate the internal stress of the cold-rolled steel sheet and transform it into steel with good workability, an annealing process is necessary. The steel sheet that has undergone the annealing process and has a recrystallized structure has a lower hardness, yield point, tensile strength, and improved workability.

In the early stage, the batch annealing method was used for annealing. Japanese steelworks have attempted to commercialize a CAL that can further improve productivity and evenness of sheet quality. In 1959, a CAL dedicated to tin coating was built at the Hirohata Steelworks. In 1972, CAL, which integrated electrolytic cleaning to recoil lines and included heat treatment facilities to prevent over-aging in the annealing section, was introduced in Japanese steelworks. This CAL was widely appreciated for producing steel plates with uniform surface quality and shape compared to the existing batch annealing furnace and was expanded to steel mills globally.[50]

The CAL is comprised of electrolytic equipment, annealing equipment, skin pass, temper rolling equipment, and other components. The cleaning facility removes contaminants attached to the surface of the strip, and annealing is performed continuously in a loop while recoiling. After annealing, the steel sheet was subjected to a light temper light temper rolling process to improve the quality and correct the shape. The recently operated CAL is shown in Figure 3.22. The pot for coating and the cooling device that forms the coating layer are connected immediately after annealing, and this process is called a hot dip continuous galvanizing line

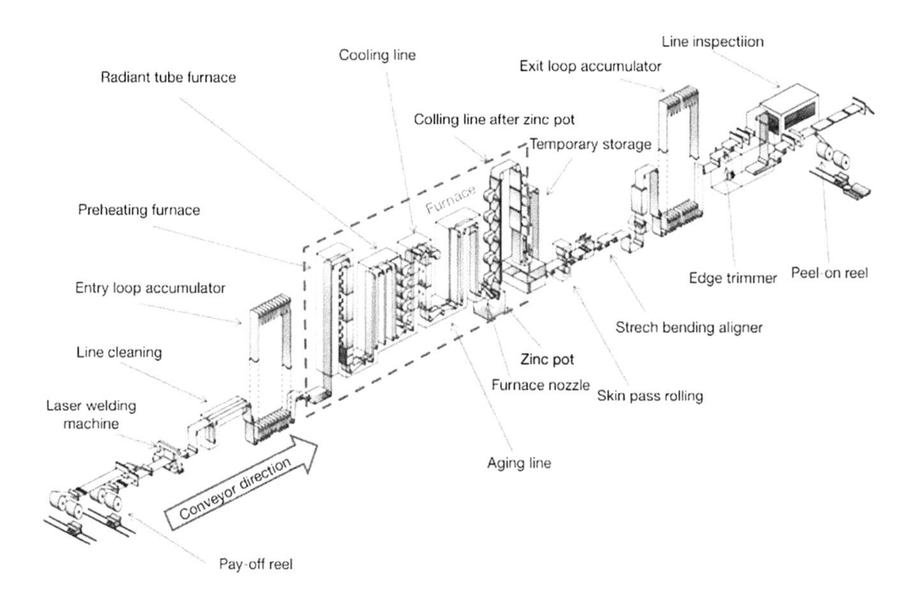

FIGURE 3.22 Typical Continuous hot dip galvanizing line. (With permission from Saltzgitter, https://www.salzgitter-flachstahl.de/en/news/details/setup-of-a-third-hot-dip-galvanizing-line-12422.html)

(CGL).[51] As a result, the CGL produces not only solid solution-hardening steel but also high-strength steel sheets of transformation-hardened steel.[52]

Since the late 1970s, customers in North America and other regions have been continuously demanding improved corrosion durability for automobile bodies. In response to this demand, steel mills worldwide have attempted to continuously produce zinc-coated steel sheets with galvanic protection. In Europe, the galvanized (GI) coating method was adopted, which directly cools the steel sheet that has passed through the zinc pot to form a coating layer. European automakers could use this steel sheet without major issues. However, Japanese automakers that exported many products to European automakers faced many difficulties in welding and painting the body when using GI steel plates. To address these issues, Japanese automobile companies collaborated with local steel mills to develop a new galva annealing (GA) process.[52]

The GA process heats the steel sheet passing through the zinc pot of the existing CGL, allowing the zinc in the base material to diffuse into the base material to form a zinc-iron alloy layer. This GA-coated steel sheet has more facilities than the existing CGL but has the advantage of having sufficient corrosion resistance even with a thinner coating thickness compared to GI. It satisfies all of the press formability, weldability, and paintability requirements of Japanese automobile companies. However, steelworks in Europe are hesitant to produce this GA-coated steel sheet because of the narrow operating window in all processes from steelmaking to final coating.

REFERENCES

1. H. Liu, Y. Qin, Y. Yang, and Q. Zhang, "Influence of Al_2O_3 content on the melting and fluidity of blast furnace type slag with low TiO_2 content", *Hindawi J. Chem.*, Vol. 2018, Article ID 9502304, 6 pages, https://doi.org/10.1155/2018/9502304
2. H. Tomohiko and S. Yoshinari, "Hiroyuki, operation guidance technique of blast furnace using data science", *JFE Tech. Rep.*, No. 26, March 2021, https://www.jfe-steel.co.jp/en/research/report/026/pdf/026-05.pdf
3. S. Shaik, S. Bagchi, S. Ghosh, and A. K. Mukherji, "Beneficiation of iron ore fines to improve blast furnace productivity", Proceedings of the XI International Seminar on Mineral Processing Technology (MPT-2010), NML, Jamshedpur, India, December 2010, https://eprints.nmlindia.org/2405/
4. "Steel | Etymology, origin and meaning of steel by etymonline", https://www.etymonline.com/word/steel
5. F. C. Thompson, "Réeaumur on steel", *Nature*, Vol. 180, No. 4597, 1957, page 1263, https://www.nature.com/articles/1801263a0
6. "Réaumur discovers carbon's role in hardening steel summary", Last updated on November 10, 2022, https://wikisummaries.org/reaumur-discovers-carbons-role-in-hardening-steel/#google_vignette
7. J. P. Schotsmans, "A pioneer of experimental metallurgy: Monsieur de Réaumur", *JHMS*, Vol. 24, No. 2, 1991, https://hmsjournal.org/index.php/home/article/view/567
8. R. F. Tylecote, "A History of Metallurgy", 2nd ed., The Institute of Materials, 1992, p. 56, https://www.academia.edu/40301278/HISTORY_OF_METALLURGY_2nd_Edition
9. H. R. Schubert, "History of the British Iron and Steel Industry from 450 B.C. to A.D. 1775", London and Beccles: William Clowes and Sons, Limited, 1955, p. 57, https://archive.org/details/in.ernet.dli.2015.104051
10. Donald B. Wagner, "The earliest use of iron in China", Published in *Metals in Antiquity*, edited by Suzanne M. M. Young, A. Mark Pollard, Paul Budd, and Robert A. Ixer (BAR international series, 792), Oxford: Archaeopress, 1999, pp. 1–9, http://donwagner.dk/EARFE/EARFE.html
11. "Iron making history in China", http://en.chinaculture.org/library/2008-02/01/content_26524.htm
12. "Blast furnace", last edited March 11, 2024, https://en.wikipedia.org/wiki/Blast_furnace

13. Helén A. Pettersson, "From bloomery furnace to blast furnace archeometallurgical analysis of medieval iron objects from Sigtuna and Lapphyttan, Sverige", Examensarbete Inom Teknik, Grundnivå, 15 Hp, Stockholm, Sverige, KTH, 2019, http://www.diva-portal.org/smash/get/diva2:1332148/FULLTEXT02
14. R. F. Tylecote, "A History of Metallurgy", 2nd ed., The Institute of Materials, 1992, p. 76, https://www.academia.edu/40301278/HISTORY_OF_METALLURGY_2nd_Edition
15. G. Magnusson, "Lapphyttan - An example of medieval iron production", in *Medieval Iron in Society*, edited by N. Bjorkenstam et al., Stockholm: Jernkontorets Forskning, H 34, 1985, pp. 21–60
16. I. Serning, HansHagfeldt, and P. Kresten, *Vinarhyttan*, Stockholm: Jemkontorets Forskning, H 21, 1982
17. Brian G. Awty, "The development and dissemination of the Walloon method of ironworking, technology and culture", Vol. 48, No. 4, October 2007, pp. 783–803, https://www.jstor.org/stable/40061328
18. "Abraham Darby I", last edited January 19, 2024, https://en.wikipedia.org/wiki/Abraham_Darby_I
19. "Abraham Darby I", https://d3hgrlq6yacptf.cloudfront.net/5f414b8e8f9bb/content/pages/documents/pdf-abraham-darby-1-2-and-3.pdf
20. Kenneth Charles Barraclough, "The development of the early steelmaking processes", https://etheses.whiterose.ac.uk/14433/1/237901_vol.1.pdf
21. "Military history of the Northern and Southern dynasties", last edited February 29, 2024, https://en.wikipedia.org/wiki/Military_history_of_the_Northern_and_Southern_dynasties
22. Y. Li, C. Ma, Y. Murakami, Z. Zhou, Y. Yang, and Y. Li, "Cast iron smelting and fining: An iron smelting site of the eastern Han dynasty in Xuxiebian, Sichuan province, China", *Sungkyun J. East Asian Studies*, Vol. 19, No. 1, 2019, pp. 91–111, https://www.muse.jhu.edu/article/725769
23. "Steel's renaissance man – Benjamin Huntsman", https://www.westyorkssteel.com/blog/steels-renaissance-man-benjamin-huntsman/
24. "The brief history of steel in Sheffield", https://www.dhscaffoldservices.co.uk/the-brief-history-of-steel-in-sheffield/
25. "Henry cort", last edited October 7, 2023, https://en.wikipedia.org/wiki/Henry_Cort
26. "Krupp gun", last edited March 13, 2024, https://en.wikipedia.org/wiki/Krupp_gun
27. "Krupp", last edited March 20, 2024, https://en.wikipedia.org/wiki/Krupp
28. "Open hearth furnace", last edited March 7, 2024, https://en.wikipedia.org/wiki/Open_hearth_furnace
29. "James Beaumont Neilson", last edited August 21, 2023, https://en.wikipedia.org/wiki/James_Beaumont_Neilson
30. "James Nasmyth", https://electricscotland.com/history/other/nasmyth_james.htm
31. R. F. Tylecote, "A History of Metallurgy", 2nd ed., The Institute of Materials, 1992, p. 166, https://www.academia.edu/40301278/HISTORY_OF_METALLURGY_2nd_Edition

32. "The Bessemer process: What it is and how it changed history", https://dozr.com/blog/bessemer-process
33. "Gilchrist–Thomas process", last edited November 30, 2023, https://en.wikipedia.org/wiki/Gilchrist%E2%80%93Thomas_process
34. "The story of the Linz-Donawitz process", https://www.voestalpine.com/group/static/sites/group/.downloads/en/press/2012-broschuere-the-linz-donawitz-process.pdf
35. R. F. Tylecote, "A History of Metallurgy", 2nd ed., The Institute of Materials, 1992, p. 136, https://www.academia.edu/40301278/HISTORY_OF_METALLURGY_2nd_Edition
36. "Induction Furnaces – A short historical notes", https://www.insertec-store.com/wp/blog/induction-furnaces-a-short-historical-notes/
37. "Steel statistical yearbook 2019, concise version", https://worldsteel.org/publications/bookshop/?filter_publication-subject=steel-data-and-statistics
38. "Heroult electric arc furnace", https://www.hmdb.org/m.asp?m=70327
39. C. O. Kang, "Revisiting the history of steel production process and its future direction", Special Report, *POSRI*, Vol. 1, January 2016, https://www.posri.re.kr/files/file_pdf/59/34/6460/59_34_6460_file_pdf_1453856769.pdf
40. UK Patent No. 221, April 3, 1857
41. Ali Maleki, Aboozar Taherizadeh, and Nazanin Hosseini, "Twin roll casting of steels: An overview", *ISIJ Int.*, Vol. 57, No. 1, 2017, pp. 1–14, https://www.researchgate.net/publication/311901275
42. US Patent 2,135,185, "Process for continuous casting of a rods", Siegfried Junghans, Willingen, Germany, Application October 12, 1934, Serial No. 48,046 in Germany October 19, 1933
43. Y. K. Shin, J. Choi et al., "Continuous casting of steel", *RIST*, Pohang, 1987, pp. 7–24 (in Korean)
44. W.R. Irving, "Continuous Casting of Steel", ISBN 9780901716538, Boca Raton: Published by CRC Press, September 1, 1993
45. S. H. Lee, "A casting speed point of view in POSCO", *BHM Berg- und Hüttenmännische Monatshefte*, Vol. 163, No. 1, January 2018, pp. 3–10, doi:10.1007/s00501-017-0695-3
46. Peter Campbell, Gerry Gillen, Wal Blejde and Rama Mahapatra, "The CASTRIP process for twin-roll casting of steel start-up experience at NUCOR'S Crawfordsville plant", 4th European Continuous Casting Conference Birmingham, October 14–16, 2002, pp. 882–890
47. "Classification of Rolling mills", Classification of Rolling mills – IspatGuru
48. Kenji Horii, Toshihiro Usugi, and Hideaki Furumoto, "Development of a bar-joining apparatus for endless hot rolling", *Mitsubishi Heavy Ind. Tech. Rev.*, Vol. 46 No. 3, September 2009, https://www.mhi.co.jp/technology/review/pdf/e463/e463029.pdf
49. "POSCO more than now, continuous hot rolling", https://www.posco.co.kr/homepage/docs/eng6/jsp/dn/company/archive/2011_eng_brochure.pdf
50. P. R Mould, "An overview of continuous-annealing technology for steel sheet products", *JOM*, Vol. 34, 1982, pp. 18–28, https://doi.org/10.1007/BF03339145

51. G. W. Bush, "Developments in the continuous galvanizing of steel", *JOM*, Vol. 41, 1989, pp. 34–36, https://doi.org/10.1007/BF03220301
52. Akihiko Inoue, Masaji Aiba, Hiroshi Hanaoka, and Yoshiaki Iwamoto, "Development of processing technology for flat sheet products", *Nippon Steel Tech. Rep.*, No. 101, November 2012, https://www.nipponsteel.com/en/tech/report/nsc/pdf/NSTR101-15_tech_review-2-4.pdf

CHAPTER 4

The Influence of Steel on Civilization and Culture

In the vast expanse of the universe, no planet resembling the Earth has yet been discovered to harbor intelligent beings like humans. Iron, a vital element, resides within the core of our planet and generates a magnetic field that shields human life from the harmful effects of solar winds. Moreover, our Earth possesses a life-sustaining atmosphere where oxygen and water are present. With the advent of iron in human society, a transformative era known as the Iron Age emerged. Throughout history, nations with advanced iron and steel technologies have risen as global powers. This raises the question: how has iron influenced the development of human cultures, including their customs, ideas, and religions, which shape our intellectual landscape?

4.1 THE ORIGIN AND SPREAD OF THE IRON AGE

Humans discovered the potential of utilizing fire approximately 1 million years ago.[1] However, as population growth presented challenges due to factors like climate change and food shortages, humans sought refuge in temperate regions near rivers. These areas offered fertile land and soft soil, ideal for agriculture, leading to the emergence of the Bronze Age and the first human civilizations. Mesopotamia and the Nile River Delta, known

DOI: 10.1201/9781003419259-4

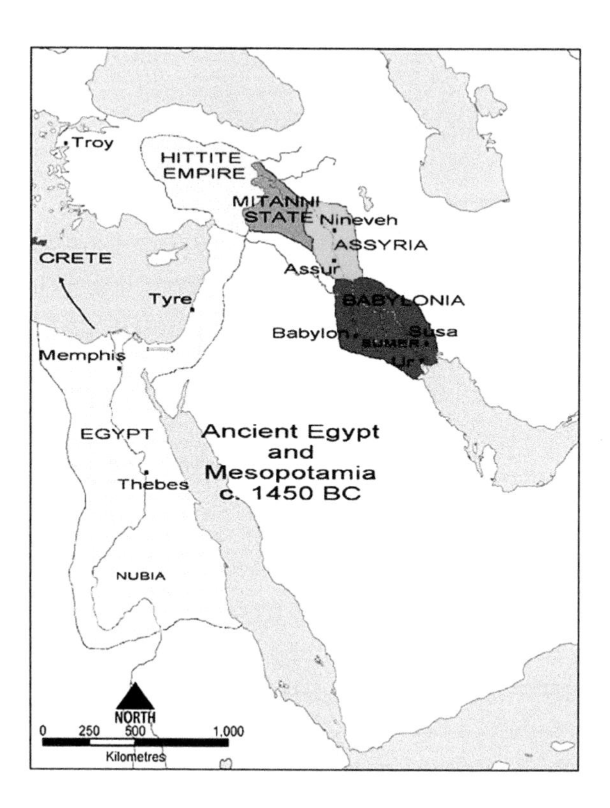

FIGURE 4.1 Fertile Crescent.[2] (Courtesy of https://commons.wikimedia.org/wiki/File:Ancient_Egypt_and_Mesopotamia_c._1450_BC.png)

as the "Fertile Crescent" (depicted in Figure 4.1), served as the cradle of human civilization. This region gave birth to various innovations such as writing, the wheel, agriculture, and irrigation, with bronze playing a pivotal role in advancing this civilization.[2]

In the quest for metals, ancient humans stumbled upon a metal that displayed a dazzling brilliance when the fire was extinguished. This marked the beginning of metal smelting. Historically, gold,[3] a favored material of the ruling class for adornments, was the first metal mankind attempted to smelt. By around 4000 BC, humans successfully smelted copper by reaching temperatures of approximately 1,000°C. Copper, with its distinctive red color, possessed excellent ductility and malleability, making it suitable for crafting jewelry and bowls. Resourceful blacksmiths soon discovered that, by alloying copper with metals such as lead, tin, or zinc, they could achieve lower smelting and casting temperatures. This new alloy, known as bronze, had a melting point approximately 150°C lower than that of pure copper, and its strength was several times greater. Rapidly, bronze wares spread as weapons, tools, and household items, ushering in the Bronze Age.

The use of iron ore 14,000 years ago in South Africa was mentioned in the literature.[4] This iron ore was not used to smelt iron but was rather used for cosmetic purposes. Later, iron production activities were documented to have taken place in several places in Africa as early as 3000–2500 BC, such as in habitation sites such as Balimbé, Bétumé, and Bouboun; smelting sites such as Gbabiri; and forge sites such as Ôboui and Gbatoro in Africa.[5, 6] However, it has been widely believed that Anatolia is the birthplace of iron smelting, because of its clear historical evidence for artifacts of smelting, weapons of iron and war records.

4.1.1 Anatolia: The Birthplace of Iron Smelting

The Hittites, an empire that reigned over Anatolia from the 18th to 13th centuries BC, were the pioneers of the world's first iron civilization. They dominated neighboring nations with their employment of iron weapons and chariots reinforced with iron. Hattusa, situated on the elevated plains of Anatolia, approximately 150 km east of Ankara, the capital of Türkiye, was chosen as the Hittite Empire's capital by Hattushili I around 1650 BC. Despite its challenging environment, characterized by cliffs, mountain peaks, extreme temperature variations, and strong highland winds, Hattusa proved ideal for habitation due to the abundant presence of iron ore in the area and favorable climatic conditions for iron smelting.[7]

Utilizing the power of iron, the Hittites conquered the neighboring Mitanni Empire and engaged in territorial conflicts with Egypt.[8] However, like other Mediterranean coastal states, the mighty Hittites fell around 1200 BC. Nevertheless, their iron smelting technology was disseminated to other regions.[9] In 2005, numerous iron artifacts, such as arrowheads and daggers, were discovered in the ruins of Kaman–Kalehöyük in Anatolia (Figure 4.2).[10] These daggers were crafted through mankind's earliest smelting of iron ore between 2100 and 1950 BC. Iron smelting also took place in nearby Georgia and Armenia, indicating the prevalence of early iron civilization in the broader Anatolian region.

Anatolia stood at the forefront of metal smelting technology, with bronzeware production also flourishing in this area. The Hittites, settling in this region, frequently clashed with Babylonia, a powerful nation in Mesopotamia. They discovered how to effectively employ iron as a weapon to defeat their adversaries. Using bloomery furnaces, the Hittites

FIGURE 4.2 A photograph of the Kaman–Kalehöyük excavation site. (Credit Hajo-Muc. Courtesy of https://commons.wikimedia.org/wiki/Category:Kaman-Kaleh%C3%B6y%C3%BCk#/media/File:Kaman-Kaleh%C3%B6y%C3%BCk1.JPG)

produced steel through the processes of smelting, forging, and carburizing iron. Hittite literature reveals the utilization of this steel in crafting helmets (depicted in Figure 4.3), as well as swords, spears, shields, and chariots.[8] Unfortunately, no artifacts related to swords or chariots have been excavated thus far.

While bronze could be smelted at temperatures around 1,000°C, iron required significantly higher temperatures of 1,500°C or above for smelting and slag separation. Achieving such high temperatures was unattainable during that era. In Hittite, wrought iron underwent repeated heating at high temperatures, accompanied by hammering to eliminate impurities. It was then reheated with charcoal and subjected to carburization. This process, known as cementation (as presented in Chapter 2 and Table 3.1), resulted in the production of steel containing 0.2 to 2% carbon.

4.1.2 Iron, the Core of Greek Civilization and Culture

The collapse of the Mycenaean civilization in Greece and the Hittite Empire in Anatolia around 1200 BC was marked by severe droughts, earthquakes,

FIGURE 4.3 Hittite iron artifacts: helmet (left) and chariot (right). (Left, courtesy of https://www.livius.org/pictures/turkey/bogazkale-hattusa/hattusa-museum-pieces/bogazkale-early-iron-age-helmet/, right, courtesy of https://commons.wikimedia.org/wiki/File:Museum_of_Anatolian_Civilizations091.jpg)

famines, and invasions by the Sea Peoples in the Mediterranean region. Cities were destroyed, trade routes vanished, and a Dark Age descended upon the land. The Hittites, possessing advanced iron manufacturing technologies, dispersed to different regions. As the aristocratic classes crumbled, written records were lost, and only a few remnants remained. The population drastically declined due to widespread famine.

However, this period provided an opportunity for Hittite-exclusive iron smelting technology to be disseminated to other regions such as Greece, Egypt, and Assyria. Obtaining tin, the primary raw material for bronze smelting, became challenging, leading to a steep rise in the cost of producing bronze.[11] Iron tools became essential for agriculture and daily life, even among common people. Blacksmiths settled in iron-rich areas and manufactured iron tools, which became more affordable than bronze tools. Farmers embraced iron tools positively, using iron axes and saws to clear forests effortlessly, expanding the arable land. The timber acquired through this process could be transformed into charcoal for iron smelting. With more settled farmers, the population grew, and surplus time was created.

Iron had a profound impact on the development of ancient human thought and religion. German philosopher Karl Jaspers (1883–1969) introduced the concept of the "Achsenzeit" or "Axis Age" in his work, "Vom Ursprung und Ziel der Geschichte."[12] From the 8th to the 3rd century BC, during this Axis Age, new religious and philosophical ideas emerged, such as Zoroaster in Persia, Buddha in India, Confucius in China, and Socrates in Greece. This Axis Age coincided with the Iron Age, which spurred the agricultural revolution, ensuring sufficient food production and generating a surplus that influenced various aspects of human life, fostering enlightenment.

Around 900 BC, the Greeks became prosperous enough to preserve precious iron items like swords, spears, and daggers, thanks to economic changes driven by agriculture and trade. This marked the earnest beginning of the Iron Age. The Greeks came into contact with the Phoenicians across the Mediterranean and adopted the Phoenician alphabet for their written language. Between 750 and 500 BC, the Greeks were able to transcribe their rich oral literature using this script, including the two famous epics attributed to Homer.

There are abundant written records that attest to the significance of iron in ancient Greece and its profound impact on their culture. Homer's *Iliad* and *Odyssey*, celebrated masterpieces of ancient literature completed around 880 BC, contain numerous references to iron farming tools and the trade associated with them.[13] Iron is also extensively mentioned in Herodotus' *Histories*, written in 446 BC, often referred to as the "Father of History."[14] Aristotle, in 350 BC, makes mention of the iron mines of Elba[15] as well as the mines of Chalybes in the Hittite Empire.[16]

During the Greek Classical period, spanning from 500 to 323 BC, cultural achievements reached their zenith, and notable figures such as Socrates (470–399 BC) emerged as influential philosophers.[9] This era also witnessed the establishment of a new form of governance known as democracy, which entailed rule by the people. As the population of Greek city-states, or Poleis, soared, the need for colonization became apparent. The most powerful Polis of the time, Athens, took the lead in colonization efforts and relocated numerous inhabitants. As depicted in Figure 4.4, dozens of colonies were established in Anatolia, North Africa, and the Mediterranean Sea, with approximately 40% of Greeks living in colonies by 500 BC.[9] It was during this period, particularly in the trading port of Miletus by the Ionians on the west coast of Asia, that a new intellectual movement known as philosophy emerged.

Iron, a material once attributed solely to the gods, was now being smelted by humans, as evidenced by the artifacts depicted in Figure 4.5.

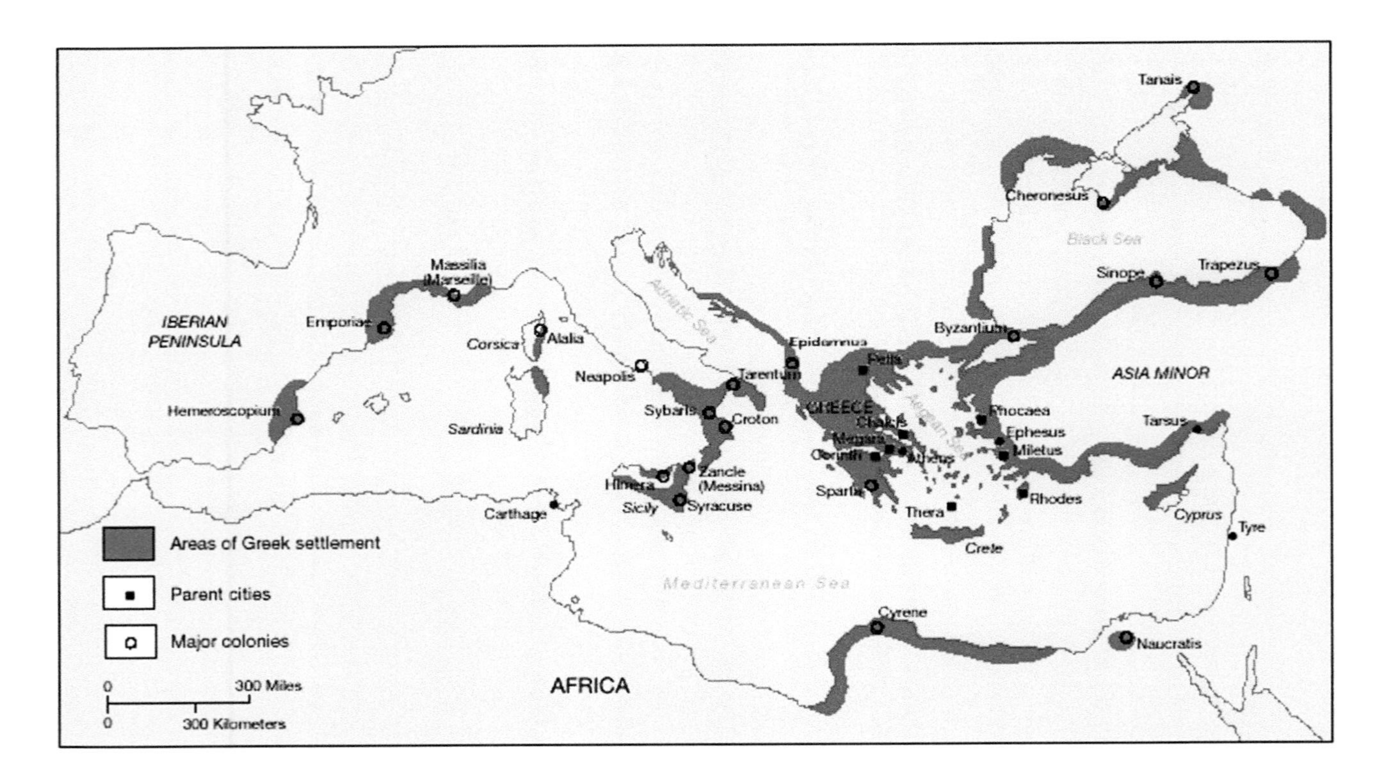

FIGURE 4.4 Greek colonization in the Archaic period. Areas in red show the colonies. (Courtesy of https://commons.wikimedia.org/wiki/File:Greek_Colonization_Archaic_Period.png)

FIGURE 4.5 Greek iron artifacts for agriculture (left) and bloomery smelting carved forges on pottery (right) (~500 BC). (Left, courtesy of https://www.khanacademy.org/humanities/whp-origins/era-3-cities-societies-and-empires-6000-bce-to-700-c-e/32-long-distance-trade-betaa/a/read-the-iron-age-beta; right, with permission from British Museum, https://www.britishmuseum.org/collection/object/G_1846-0629-45)

The availability of iron led to an agricultural surplus, facilitating increased trade and interactions with foreigners. These social changes fostered the emergence of rational thinking grounded in logic and science. Philosophers of the time, such as Thales of Miletus, often referred to as the grandfather of philosophy, pondered the existence of a fundamental substance underlying natural phenomena. Thales posited that it was "water," while Anaximander proposed the concept of the "infinite," and Anaximenes suggested "air." Pythagoras explored the realm of "number," Empedocles put forth "water," "fire," "air," and "earth," Anaxagoras posited "seeds," and Democritus argued that all things were composed of indivisible "atoms."(17) Socrates, an Athenian philosopher who ushered in a new era of philosophical inquiry, regarded people's thoughts and opinions as the starting point of philosophy. Plato (4th-century BC) centered his philosophy around the concept of "ideas," while Aristotle (384-322 BC, Figure 4.6) favored an empirical approach grounded in observations and experiences.(18) These philosophical developments had a profound impact on the intellectual landscape of ancient Greece during the Classical period.

At the age of thirteen, Alexander the Great began his studies in natural sciences and humanities under the tutelage of Aristotle. He ascended

FIGURE 4.6 Plato (left) and Aristotle (right) pointed to heaven (ideal) and earth (reality) respectively. (Courtesy of https://en.wikipedia.org/wiki/The_School_of_Athens#/media/File:%22The_School_of_Athens%22_by_Raffaello_Sanzio_da_Urbino.jpg)

the throne in 336 BC and embarked on an eastern expedition in 334 BC. During his conquests, he dismantled the Achaemenid Persian Empire and established an expansive dominion that extended across Central Asia and northwestern India. Throughout this campaign, Alexander placed significant emphasis on the preservation and handling of iron. However, his intentions were not directed toward weapon production but rather the refinement of jewelry. Alexander's specific instructions to his generals were focused on the polishing of diamonds rather than the forging of arms. During this era, the use of fine diamond powder as an abrasive agent was employed in the process of rotating rough diamonds against a metal disk to achieve a lustrous finish. Iron served as the ideal metal to securely hold the abrasive in place, preventing its escape during the polishing procedure.[19]

4.1.3 Iron, the Power of the Roman Empire

Building upon the foundation of the Greek iron culture, the Roman Empire emerged as a dominant force, exerting its influence over the entire Mediterranean region, northern Africa, and Mesopotamia, as depicted in Figure 4.7. The Roman Empire's expansion and the spread of its culture, rooted in iron, reached as far as Gaul through Caesar's conquests, effectively encompassing all of Europe. The Iron Age extended its reach from Central Asia and China to Southeast Asia, Korea, and Japan.[20] The inhabitants of ancient Rome, residing in central Italy, possessed knowledge of metal smelting dating back to the Bronze Age. However, the central region of Italy itself was not abundant in iron ore. As the Roman Empire expanded, iron ore was sourced from the islands of Elba and Sardinia, situated to the north and west of the Italian peninsula. Additionally, through their victory in the Punic Wars (264–146 BC) against Carthage, the Roman Empire gained control over Gallia Narbonensis in southern France, an area abundant in iron ore reserves. Figure 4.8 illustrates the key regions where iron production thrived during the reign of the Roman Empire.[21]

Elba Island, famously known for Napoleon's exile, held great significance as a center for iron smelting within the Roman Empire. Archaeological evidence reveals traces of iron smelting on the island dating back to the 5th century BC. However, iron smelting on Elba abruptly ceased during the middle of the 1st century BC. A persuasive hypothesis suggests that the cessation of iron smelting was due to a shortage of wood for fueling the smelting furnaces. It appears that the emission of soot from the iron smelting process on Elba Island was substantial during that time. The ancient Greek name for "Elba" is "Aithále," which translates to "sooty."[22]

The Colosseum, a magnificent structure symbolizing Rome, was constructed in the 1st century AD and could accommodate over 5,000 spectators. During the construction of the Colosseum, approximately 300 tons of iron were utilized to secure the stones, as depicted in Figure 4.9.[23] Iron clamping, a common method employed in large Roman buildings, was utilized, and the surfaces were coated with lead to prevent corrosion.

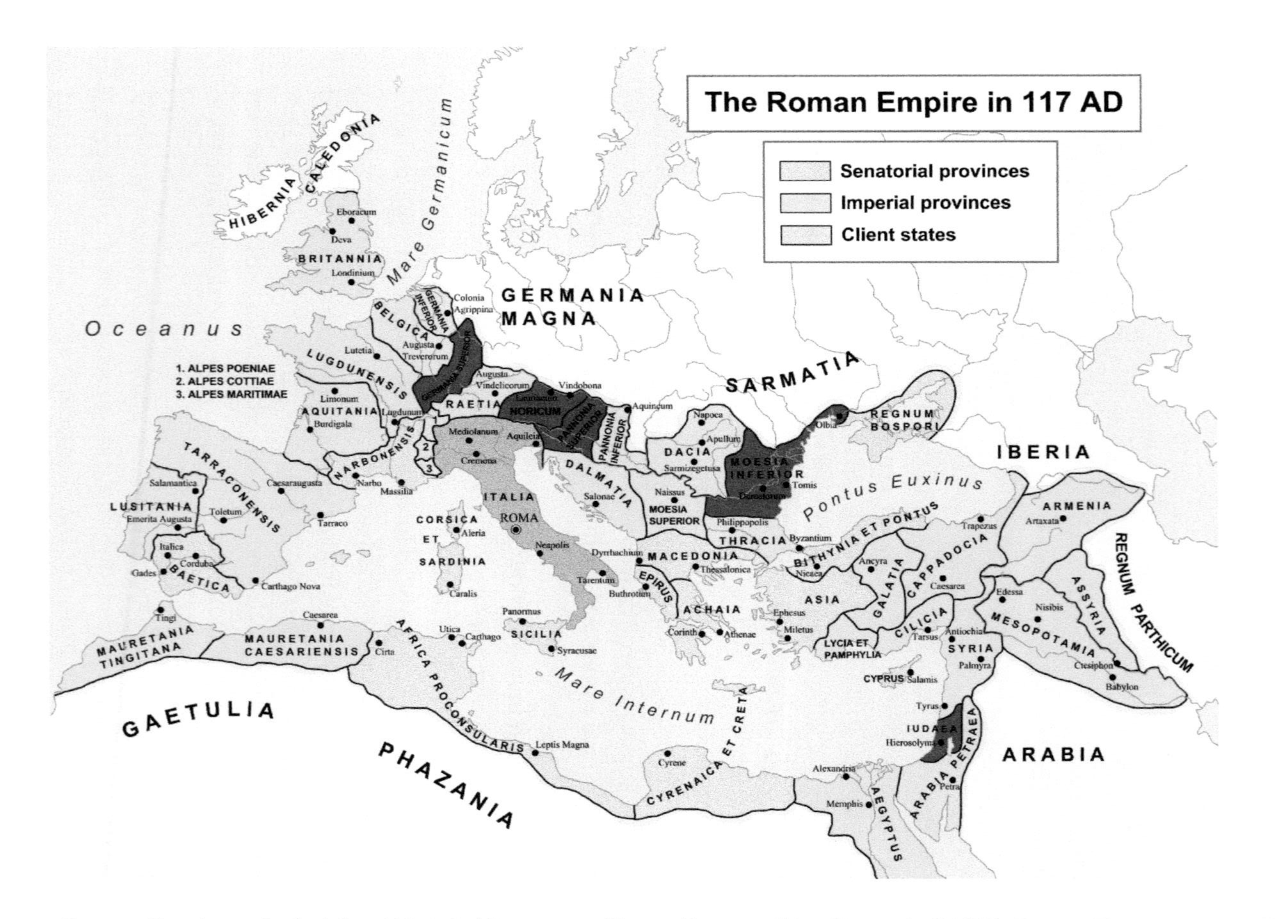

FIGURE 4.7 Roman Empire at its height, AD 117. (Courtesy of https://en.m.wikipedia.org/wiki/File:RomanEmpire_117_-_Earliest_locations_of_Mithraism.svg)

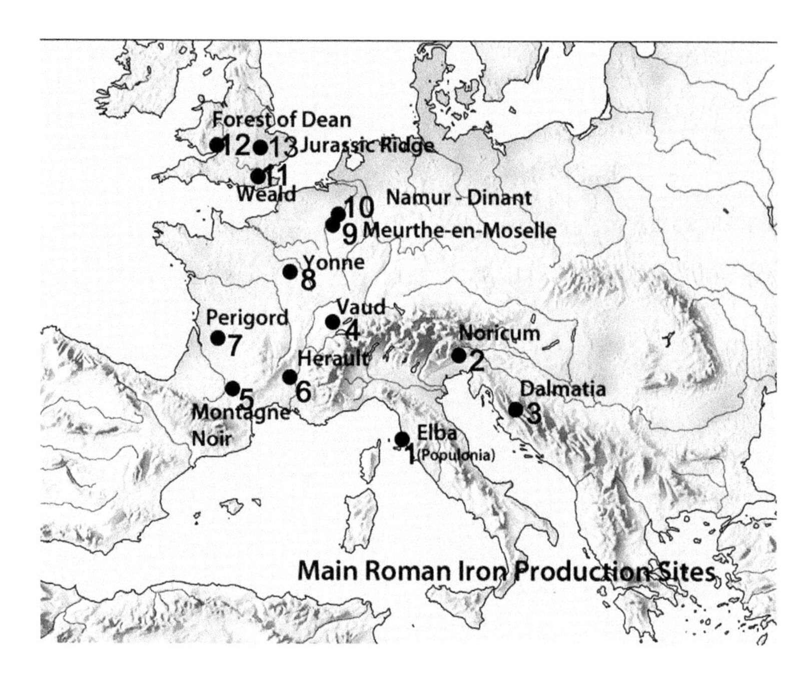

FIGURE 4.8 Major iron production regions of the Roman Empire.[21] (With permission from Taylor & Francis, chrome-extension://efaidnbmnnnibpcajpcglclefindmkaj/https://www.tf.uni-kiel.de/matwis/amat/iss/kap_a/articles/2017_janet_lang_roman_iron.pdf)

Material	Usage	Quantity (if known)
Cement/Concrete	Foundation, upper floors, vaulted arches	250,000 cubic meters
Travertine (limestone)	Main pillars, ground floor, external wall	100,000 cubic meters
Volcanic tuff (peperino)	Minor pillars, radial walls/skeleton	
Iron	Clamps for binding travertine	300 tons
Tiles	Floors and walls	
Bricks	Walls	
Marble	Seating, statues and ornaments, drinking fountains, covering for outside walls	
Lead and terra-cotta pipes	Water and Sewer system	

FIGURE 4.9 Iron clamps holes in the exterior of the Colosseum (left) and amount of iron used to the Colosseum.[23] (Left, with permission from Megan Anderson, https://engineeringrome.org/ancient-structures-in-rome-the-colosseum-pantheon/; right, courtesy of https://www.semanticscholar.org/paper/A-Computer-Generated-Model-of-the-Construction-of-Tan/38e8ade1621ece391e57c97dec762c8947fabc02#extracted)

The Gladius, a renowned sword of the Roman Empire, is depicted in Figure 4.10, showcasing the Tiberius Gladius from the first half of the 1st century AD, currently housed in the British Museum. The blade's edge underwent heat treatment, transforming it into a martensite structure.

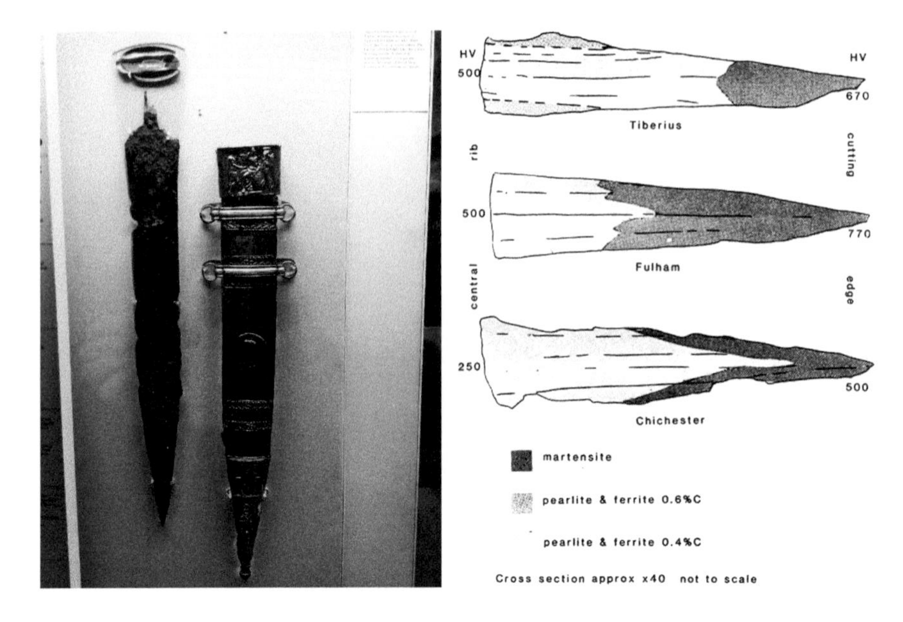

FIGURE 4.10 Tiberius Gladius was created in the Roman Empire in the first half of the 1st century AD. Left: Tiberius appearance, right: composition and hardness distribution of Tiberius and the other two Gladiuses as a function of their location.(24) (left, courtesy of Wikimedia Commons, https://commons.wikimedia.org/wiki/File:Mainz_sword.jpg; right, with permission from Copy Clearance Center, https://www.jstor.org/stable/526199)

Considering the variation in carbon content between the blade and the core, it can be inferred that different carburization techniques were employed for each component.(24) The Gladius will be further discussed in Chapter 6.

There is an ongoing debate regarding the iron production of the Roman Empire during its peak, with some arguing that it far exceeded that of other regions at the time. According to Irish ancient historian Raoul McLaughlin, the iron production of the Roman Empire reached approximately 80,000 tons per year, surpassing the production of other contemporary civilizations, including China, which was claimed to be sixteen times less.(25) However, this assertion contradicts the arguments put forth by other scholars.(26) Nevertheless, both the Roman Empire in the West and the dominant powers in the East placed significant emphasis on iron production, as it was crucial for agricultural tools and essential in warfare.

Kevin Greene argues that the economic prosperity of the Roman Empire can be attributed to its geographic, political, and social integration. This unprecedented level of integration allowed the empire to access and exploit resources more effectively, as well as to capitalize on trade opportunities.[27] The Roman Empire's extensive reach and interconnectedness, both geographically and socially, contributed to its economic success.

4.1.4 Iron, the Root of Indian Culture

Recent investigations in India have uncovered several iron artifacts dating back to 1500 to 1000 BC, shedding light on the significance of iron in Indian culture.[28] During the Bronze and Iron Ages, trade and warfare connected the continents, spanning from the eastern Mediterranean to western India. Around 1500 BC, the nomadic Aryans invaded the Indian subcontinent and settled along the Ganges River. With their knowledge of iron smelting, the Aryans transformed the dense forests of the Ganges Valley into fertile farmland using iron axes. The warm and humid environment of the Ganges Valley proved ideal for agriculture, leading to the establishment of densely populated settlements and the expansion of agricultural land throughout the Indian subcontinent, marking the onset of the Iron Age.

The Aryans also introduced a caste system and a distinct religion to differentiate themselves from the native inhabitants. This caste system became deeply ingrained in Indian society and has ancient origins. It evolved over time and was influenced by various social, religious, and cultural factors in India. Brahmanism, an early form of Hinduism, played a role in shaping some aspects of the caste system. Despite the rapid agricultural development and population growth, the lower classes engaged in agriculture continued to struggle economically, unable to escape their designated social classes. In the fifth century BC, Siddhartha Gautama, later known as the Buddha, advocated for mercy and equality, resonating with many lower-class individuals who embraced Buddhism. To accommodate this growing movement, the upper-class Brahmans transformed the existing Brahmanism into a new religion called Hinduism, which incorporated various religious beliefs, including Buddhism.[29]

Iron played a crucial role in the modernization of India. One notable example is the iron pillar in Delhi, the capital of India, which weighs 7 tons and was created in AD 310. As depicted in Figure 4.11, this pillar stands tall and remains in remarkable condition. Historical records indicate that iron and steel from India were exported to various distant lands, including Rome, where they were used in the construction of world-famous bridges and monuments. In India, a prevalent iron smelting technique involved the use of crucibles. This method gave rise to a special product known as

FIGURE 4.11 Iron pillar in India. (Courtesy of https://commons.wikimedia.org/wiki/File:QtubIronPillar.JPG)

Wootz steel, which became the primary material for crafting the renowned Damascus swords that will be discussed further in Chapter 6.[30]

In 1996, scholars made an intriguing discovery linking the production of high-quality steel in Sri Lanka and India to the monsoons from the Indian Ocean. British and Sri Lankan archaeologists unearthed the remains of forty-one iron smelting furnaces in Sri Lanka, providing evidence of the use of monsoons in the production of high-carbon steel. Through the replication of ancient furnaces, researchers were able to recreate the conditions that utilized natural wind pressure and charcoal heat to manufacture the exceptional Wootz steel.[31]

4.1.5 Iron, the Birth of Confucianism in China

According to Dr. Donald Wagner's hypothesis, iron smelting technology was introduced to China by the Scythian nomads in Central Asia who

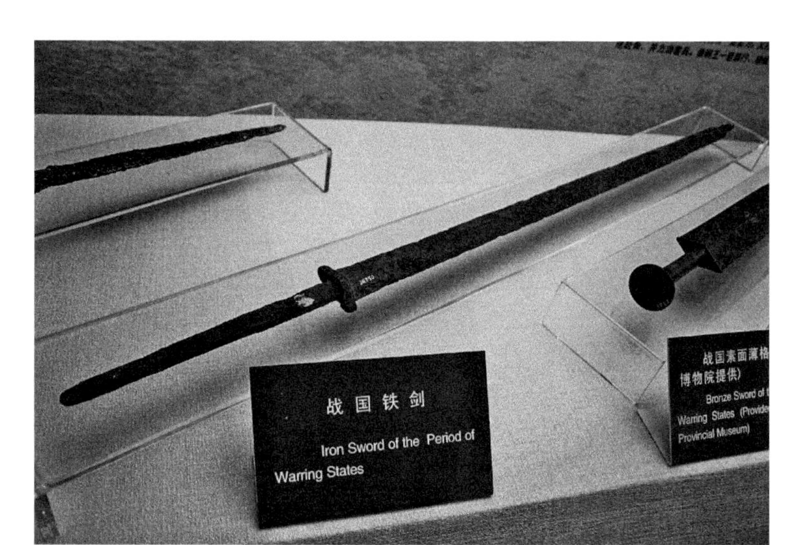

FIGURE 4.12 Iron sword of the period of Warring States. (Courtesy of Wikimedia Commons, https://en.wikipedia.org/wiki/Chinese_sword#/media/File:Warring_States_Iron_Sword.jpg)

employed bloomery smelting techniques during the 8th century BC.(32) Evidence suggests that as early as the 11th century BC, China began manufacturing daggers with meteorite inserts in bronze handles, and by the 5th century BC, iron blades were being produced through smelting.(32) However, Chen J. reported that one of the two iron artifacts excavated in Gansu in Central China was smelted in a bloomery furnace.(33) So, it is still a matter of debate whether Chinese iron smelting technology was diffused from outside or self-developed.

The discovery of a group burial site from the Western Zhou Dynasty near Sanmenxia, Henan Province, China, provides further insight into the beginning of China's Iron Age civilization, particularly in the northwest region of the continent.(34) Among the artifacts excavated from this site, five iron objects were found, three of which were made from meteorites and two from smelted iron. By the end of the third century BC, the Chinese had learned how to produce quench-hardened steel swords. Figure 4.12 shows an example of a long sword made of steel during the period of Warring States Period.

China's Iron Age began approximately 500 years later than that of West Asia; however, during the 4th century BC, an innovative technology for producing molten iron with a carbon content of over 2% and a melting point of around 1,150°C was introduced through blast furnaces.(35) This breakthrough allowed for the production of exceptionally hard and

FIGURE 4.13 Chinese ancient casting vessel. The legs of this vessel are cast iron, while the body is bronze.[36] (Courtesy of https://commons.wikimedia.org/wiki/File:Warring_States_Bronze_Ding_03.jpg)

corrosion-resistant castings. However, these castings were brittle and required additional heat treatment. Ancient Chinese blacksmiths mastered the technique known as the co-fusion method, enabling them to achieve the desired characteristics in their castings. Figure 4.13 displays a casting product created using this method.[36]

After the Warring States Period (403–221 BC), cast iron emerged as a significant material for the production of household items and civil engineering tools. During the Han Dynasty (221 BC~AD 220), approximately 36,000 tons of cast iron were estimated to have been produced.[26] The Han Dynasty experienced a period of prosperity by bolstering agricultural output and achieving success in warfare. Iron became indispensable for efficient governance, leading to its inclusion, along with salt, in a monopoly system known as the Policy of Salt and Iron, or 塩鐵論.[37]

During the Spring and Autumn Warring States Period in China, iron tools were supplied for agricultural purposes, allowing commoners to cultivate and harvest crops in state-granted farmland. A portion of the harvested crops was contributed to the state, while the remainder sustained their livelihoods. Commoners fulfilled dual roles as both soldiers and peasants during this era. They dedicated themselves to farming during the busy agricultural season and were conscripted for military service or civil engineering projects during the off-farming season, in response to the state's call. From the perspective of the feudal lords who governed the land, bountiful harvests were essential for collecting taxes and recruiting soldiers and skilled individuals with high morale. Agricultural productivity has become more crucial than ever before. Although bronze was valuable, it lacked the strength required for use as a farming tool. Conversely, iron ore was abundantly available across various regions. With the active support of feudal lords, mass production of iron castings became achievable, leading to a rapid transition from bronze to iron for agricultural tools and weapons. Standardization was also implemented in the iron smelting and casting processes.[38, 39]

During the Spring and Autumn Warring States Period, the increasing frequency of conscripting commoners for war and civil engineering projects led to a rise in their discontent. To address this issue, Confucianism, as advocated by Confucius (551–479 BC), emerged as a unifying ideology that appealed to both commoners and officials, providing a rational basis for their loyalty to the state and king. Confucianism emphasized the importance of aligning one's allegiance to the monarch, akin to the reverence shown toward ancestors based on human nature. Recognizing the need for the loyalty of commoners and vassals, the king readily embraced Confucius' philosophy as a framework for governing the state. The emergence of wealthy and powerful landlords, who greatly increased agricultural yields through the use of iron farming tools, burdened the king, compelling him to endorse Confucian principles that demanded loyalty from all individuals. This support further solidified the kingship. Confucius recommended that feudal lords adhere to the principle of "the King like King, Vassal like Vassal, Father like Father, Son like Son" (君君臣臣父父子子 in Chinese).[40] Although Confucius passed away before the introduction of the Iron Age in China, his disciples compiled his teachings and established Confucianism, which had a profound impact on the Chinese mindset. The Han Dynasty, renowned for its flourishing iron civilization, became the first state to adopt Confucianism as a national policy.[41]

4.1.6 Culture of the Iron Age in Korea

The culture of the Iron Age in Korea was largely influenced by its connection with China. While it is believed that ironworking knowledge was transmitted from China, there is a theory suggesting that early Iron Age relics excavated in China were passed down to northern nomadic tribes,[42] including the Gojoseon region, considered the ancestral homeland of the

Korean people. Chinese historical records indicate that ironware production and usage in Korea date back to around 100 BC. According to these records, iron was produced in the tribal countries of Byeonhan and Jinhan on the Korean Peninsula, and it was traded with the Han(韓), Yeh(濊), and Wae(倭) kingdoms, as well as being used as currency. Notably, the culture of the ancient Iron Age in Korea is closely associated with the three kingdoms that dominated the Korean Peninsula during this period: Goguryeo, Baekje, and Silla.

4.1.6.1 Goguryeo

Iron played a crucial role in the establishment and success of Goguryeo, the kingdom that occupied the largest territory in the history of the Korean Peninsula. The abundance of iron resources in the region where Goguryeo settled is believed to have contributed to the founder's ability to unite various groups.[43] According to Chinese historical records such as "Buksa" and "Three Kingdoms," Goguryeo also engaged in iron trade with the Khitan in the north.[43] The military power of Goguryeo was formidable enough to successfully resist the Sui and Tang forces that sought to unify the Chinese continent during that time.

Figure 4.14 showcases a procession from the Tonggu Tomb No. 12 Mural, depicting Goguryeo's armed infantry and cavalry in formation. This artwork reveals a wide array of weapons, including swords, spears, axes, tridents, bows, and crossbows. The arrowheads were crafted from bloomery iron, with their tips forged and quenched to form a martensite

FIGURE 4.14 Iron armor (left) in the Tonggu Tomb No. 12 Mural and a god of a blacksmith (right) pictured in Ohebun No. 4 Mural of Goguryeo. (Left, courtesy of https://commons.wikimedia.org/wiki/File:Goguryeo_armor.jpg, right, courtesy of https://www.ohmynews.com/NWS_Web/Series/series_premium_pg.aspx?CNTN_CD=A0002908169)

structure. Additionally, the discovery of shellfish powder at the burial site suggests its possible use as a desulfurization agent.[44]

4.1.6.2 Baekje

Baekje, one of the ancient Korean kingdoms, is regarded as a frontrunner in steel technology. Archaeological excavations in Baekje territory have revealed the presence of nine circular smelting furnaces that were operational during the 3rd and 4th centuries, producing iron for over a century. The scale of these iron smelting furnaces, along with the technological level of the burial artifacts found, indicates that Baekje achieved an Iron Age civilization comparable to that of Goguryeo.[45]

The advanced steel technology of Baekje can be inferred from the Japanese national treasure called "Chiljido," which is preserved at Ishinokami Shrine in Japan (Figure 4.15). Historical records also mention that during the reign of King Geunchogo (AD 346–375), the Crown Prince of Baekje presented the Chiljido, along with 40 iron lumps, to a Japanese envoy.[46] The Chiljido is believed to have been crafted from Baekryunkang, a type of steel known for its co-fusion, tempering, and forging techniques, as indicated by the inscription "Chiljido of Jo Baekryunkang" engraved on

FIGURE 4.15 Chiljido, Meaning with Seven Branched Sword. (Courtesy of https://commons.wikimedia.org/wiki/File:Seven-Branched_Sword.jpg)

the sword. The early transmission of ironmaking technology to the Baekje region is thought to have originated from China's Yan Dynasty around the 3rd century BC. Ironware, cast or forged during the 1st century BC, has also been unearthed from various locations. By the 2nd century AD, the Han River region under Baekje's rule saw abundant production of agricultural tools based on forging technology.[47]

6.1.6.3 *Silla*

Silla, one of the ancient Korean kingdoms, recognized the significance of the iron industry from its early days and established a state-run factory. Mining activities took place in Dalcheon, depicted in Figure 4.16, which was located near Geumseong, the capital of Silla since the 1st century BC. The iron ore found in Dalcheon was magnetite extracted from open pit mines. Silla constructed smelting furnaces, refining furnaces, a casting area, and forging furnaces, indicating the operation of integrated steelworks.[48, 49] Interestingly, the iron ore from Dalcheon contained trace amounts of arsenic. It is noteworthy that iron artifacts discovered in major iron production areas in Japan also contained arsenic, suggesting that iron ore from Dalcheon may have been exported to Japan.[50] In 573, Silla cast an iron statue called Jangyukjonsang, one of the three treasures of Silla, which was placed in Hwangyong Temple. This statue had a total weight of 21 tons, with approximately 7 tons of iron used in its creation. Sadly, this Buddha statue was melted down during the Mongol War of

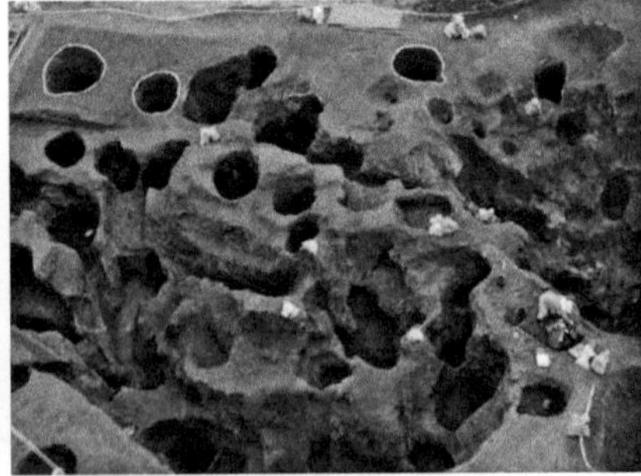

FIGURE 4.16 Excavation of Dalcheon Mine.[49] (Courtesy of https://www.cha.go.kr/cop/bbs/selectBoardArticle.do?nttId=63999&bbsId=BBSMSTR_1222&mn=NS_03_08_03&ccbaCpno=2332600400000®ionGbn=26&pageUnit=10&pageIndex=1&rnum=1)

FIGURE 4.17 Seated iron Buddha of the Unified Silla period found at the Bowonsa Temple site in Seosan. (Courtesy of https://www.museum.go.kr/site/main/relic/search/view?relicId=510)

Goryeo. However, at the time, it stood as the largest iron casting in Asia.[51, 52] Figure 4.17 showcases a 1.5 m tall Iron Buddha statue created in the 9th century during the Unified Silla Period, which is preserved at the Bowonsa Temple Site in Seosan. The abundance of iron artifacts unearthed from the royal tombs of Silla provides insight into the plentiful supply of iron during that era. This ample iron supply may have played a role in Silla's ability to unify the Three Kingdoms at the time.

4.1.7 Iron, the Origin of Japanese Craftsmanship

Iron production in Japan began in the late Yayoi period around the third century AD, and the country entered the full-fledged Iron Age during the Kofun period in the fifth century AD.[53] The knowledge of steel technology was transferred from both China and Korea. Initially, ironmaking

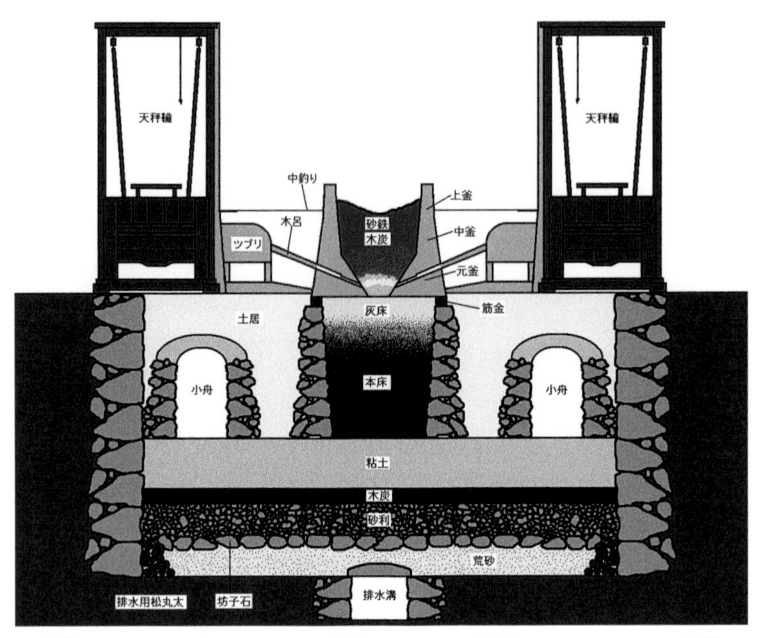

天秤鞴: bellows, ツブリ: platform, 木呂: manifold, 中釣り: medium fishing, 沙鉄: iron sand, 木炭: charcoal, 灰床: ash bed, 本床: main bed, 粘土: clay, 沙利: gravel, 上釜: upper part, 中釜: middle part, 元釜: furnace center, 節金: joint, 土居: sand floor, 小丹: clay drier, 荒沙: coarse sand, 坊子石: stone, 排水溝: drainage channel, 排水用松丸太: pine logs

FIGURE 4.18 Cross section of Tatara furnace. (Courtesy of https://commons.wikimedia.org/wiki/File:Structure_of_Eidai_Tatara.png)

in Japan was quite primitive, utilizing small kilns. However, with the arrival of migrants from China and Korea in the sixth century AD, more advanced ironmaking techniques became accessible. Skilled masters from Korean ironworks and craftsmen, following the decline of their kingdoms, Gaya and Baekje, also migrated to Japan and contributed to the further development of iron processing technology.

Japanese artisans developed their unique ironmaking method known as the Tatara ironmaking method, depicted in Figure 4.18. The majority of the approximately 150 iron relics discovered in Japan, including the Tatara iron furnace, are concentrated in Shimane Prefecture, which had convenient access to the Korean Peninsula through the East Sea current. In this region, the Tatara iron furnaces constructed by artisans are still operational to this day.(54) The term “Tatara” refers to blowing air with bellows, and the air-blowing technique is considered a crucial aspect of Japanese ironmaking technology. This is because the quality of the produced iron depends on the heating temperature, which is determined by how efficiently the air is blown through a generous amount of charcoal.(55)

The Japanese sword, known as Nihonto, holds great pride for the Japanese as a symbol of exquisite steelwork. While its purpose was originally for self-defense and combat, the Japanese sword transcended its role as a mere weapon. During the Edo period, it became a tool for self-discipline through the practice of swordsmanship and a means of social elevation. Additionally, it became an integral part of spiritual culture and a defining symbol of the samurai class. Toward the end of the Warring States period, when Toyotomi Hideyoshi held the position of the shogun, he implemented the Katanagari system. This system allowed only samurai to bear swords while confiscating them from others, including peasants.(56) The shogun entrusted renowned craftsmen with the production of Japanese swords, which he would either retain for himself or bestow as gifts upon his favored samurai.(57)

4.2 THE INFLUENCE OF IRON ON THOUGHT, ART, AND CULTURE

4.2.1 Iron, the Influential Force in European Medieval Culture

4.2.1.1 The Pillars of the European Renaissance: Iron, Ships, and Finance

During the medieval period, the trade of spices in Europe was a highly lucrative business. These valuable commodities were transported from Constantinople to Venice and sold at exorbitant prices. Venetian merchants played a significant role in this trade, actively participating in the 200-year Crusade that began in the late 11th century. By providing support to warships and establishing the world's first standing army, known as the Venetian Navy, they gained unparalleled dominance in the global trade market and amassed immense wealth. Figure 4.19 depicts the renowned Venetian Arsenal, where warships were primarily constructed.(58)

By the year 1500, the Venetian Arsenal had evolved into the world's largest complex dedicated to the production of weapons and iron products for military purposes. Situated in the eastern part of Venice and spanning approximately 3,000 m^2, this shipyard was a hub of activity. Within its premises, iron was smelted in bloomery furnaces to manufacture a wide range of essential items such as nails, iron joints, anchors, and even cannons. The galleys shown in Figure 4.20, which were cutting-edge warships of their time, relied on iron structures and components, making them

FIGURE 4.19 Entrance to the Venetian Arsenal in the 1860s.[58] (Courtesy of https://en.m.wikipedia.org/wiki/File:View_of_the_entrance_to_the_Arsenal_by_Canaletto,_1732.jpg)

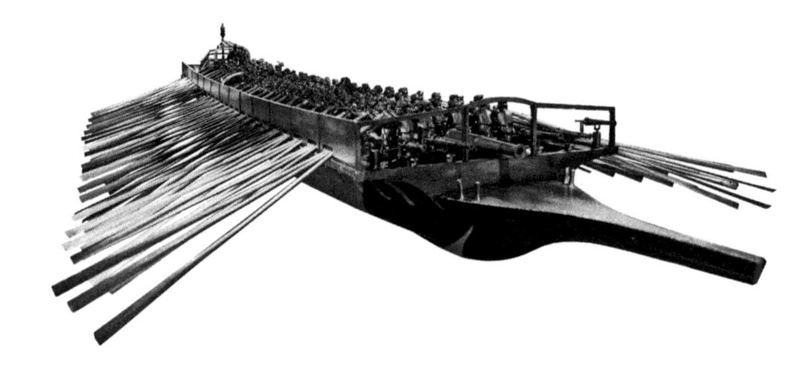

FIGURE 4.20 Venetian galley model.[59] (Courtesy of https://en.m.wikipedia.org/wiki/File:Venice_galley_rowing_alla_sensile1.jpg)

virtually invincible. At the Venetian Arsenal's drydock, a remarkable fleet of 100 galleys was always ready for action, symbolizing the impressive naval power held by Venice during that era.[59]

In the 14th century, Florence, Venice, and Genoa emerged as prominent centers of divinity. These three free cities, directly under the control of the Roman emperor, served as vital passageways connecting Europe and the markets of the Middle East, thanks to their thriving iron and

FIGURE 4.21 Florence was the center of the Renaissance movement. (Courtesy of https://commons.wikimedia.org/wiki/File:Sunset_over_florence_1.jpg)

shipbuilding industries. As the wealth of these cities grew exponentially, a financial industry began to develop, particularly in Florence, as depicted in Figure 4.21. The flourishing trade fostered a newfound interest in humanity, reminiscent of the intellectual atmosphere in Miletus, Greece. People began to appreciate the beauty of humans and explore scientific principles that had been considered taboo during the Middle Ages, thus paving the way for the Renaissance.

The Renaissance, fueled by the accumulated wealth generated by the iron industry, spread to Northern Europe, including countries like France, Germany, and England, and laid the foundation for modern European culture. Italy, in particular, witnessed remarkable advancements in art and architecture during this period. Florence, with its flourishing Renaissance movement, played a pivotal role and was home to the influential Medici family. The Medici achieved great success in commerce and finance, managing even the assets of the Holy See. Their wealth grew to such an extent that they established sixteen branches across Europe and earned the moniker 'Guardian of the Renaissance'. Leveraging their financial power, the Medici family strategically engaged with the Church of Rome and the French royal family, leading to the appointment of three popes and two French queens.[60]

4.2.1.2 Iron Sparked the Printing Revolution and the Reformation

On October 31, 1517, Martin Luther (1483–1546) posted the "Ninety-five Theses" on the church door of Wittenberg Castle in central Germany.[61] During that time, the Catholic Church was selling indulgences to finance its activities. Luther's theses served as a call for reform within the Catholic Church. His writings were printed using a metal-type press invented by Johannes Gutenberg (1397–1468). Luther's ideas quickly spread throughout Germany. Gutenberg utilized cast iron to create a type-casting machine and used a mixture of lead and tin to produce metal type.[62] Gutenberg took advantage of the fact that cast iron with a high carbon content possesses low viscosity even at low temperatures, making it suitable for casting complex shapes and ensuring durability.

Gutenberg's printing press marked a revolutionary event and is considered the most significant among the ten most important human events of the past thousand years. Before its invention, printed books were expensive and rare, primarily accessible to the aristocracy. Transcribing a single Bible took approximately two months at the time, but in 1455, the "Gutenberg Bible" required only one week to print 500 copies.[63] The advent of high-speed printing laid the groundwork for widespread distribution, greatly expanding the influence of textual information on popular culture. A printing revolution occurred, leading to an explosion of knowledge and information dissemination.

The printing revolution, triggered by iron, played a decisive role in deepening and spreading modern philosophical thinking and the spirit of scientific inquiry. It occurred during the Renaissance, a period marked by the emergence of the Enlightenment, which was based on reason and experience. The Enlightenment, a European trend of the 17th and 18th centuries, held a belief in the infinite progress of mankind and aimed to dismantle the existing order and reform society through the power of reason.[64] The ideological foundations of the Enlightenment included rationalism, empiricism, and the scientific revolution.

4.2.1.3 Iron: The Trigger of Capitalism and Socialism

The advancement of iron and steelmaking technologies, along with the Industrial Revolution it propelled, led to the rise of entrepreneurs. In 1850, the term "Capitalism" was first coined by Louis Blanc (1811–1882) of France to describe a system where individuals held exclusive ownership of the means of production, rather than it being commonly owned.[65] During the era of commercial capitalism from the 15th to the 18th centuries,

European countries adopted economic policies of mercantilism, aimed at increasing national wealth. However, in 1776, British economist Adam Smith (1723–1790) challenged the doctrinal beliefs of mercantilism with his work "The Wealth of Nations."[(66)] Smith rejected the notion that the total amount of wealth in the world was fixed and that one country could only increase its wealth at the expense of others. Instead, he recognized that wealth could be created through the use of iron machinery and factories, driven by the advancements of the Industrial Revolution and the invention of the steam engine.

The Industrial Revolution, triggered by iron, resulted in the rise of industrial capitalism, displacing commercial capital. The economic system of industrial capitalism demanded the principle of liberalism, wherein the state and any organization refrained from interfering with the production of goods and had minimal authority to prevent foreign aggression and maintain domestic order.[(67)]

The impact of the Industrial Revolution led by iron extended to various social movements, including labor movements, gender equality, urbanization, and the growth of modern consumer society. Working conditions during this period were abysmal, with factories lacking safety regulations and being prone to accidents and fires. Factory owners were not subjected to any occupational safety regulations whatsoever.[(68)]

A major social change brought about by the Industrial Revolution was the migration of people from rural to urban areas. Factory owners sought cheaper labor compared to skilled artisans, and machines required minimal skills to operate. Consequently, factory owners often employed women and children for low wages. The growth of capitalism and the pursuit of profit created a negative perception among workers regarding the exploitation of their labor power, prompting a desire for a powerful force to advocate for their rights. Around 1840, Karl Marx and Friedrich Engels developed a theoretical and practical program reflecting these changes, fueling labor-related social transformations and political revolutions.[(69)]

Marx departed from the existing "Romantic Communism" and introduced "Materialism" and the "Communist Manifesto," presenting the concept of "Scientific Communism." He characterized the history of society thus far as a history of class struggle and asserted that exploitation could only be eliminated through the abolition of social classes. Marx argued that while previous revolutions had been fought by new capitalists against feudalism, the future revolution would be led by the working class, who had suffered the consequences of the Industrial Revolution. In the

"Communist Manifesto," Marx proclaimed that the proletariat, the class that was exploited and oppressed, had reached a stage where it could no longer liberate itself from the bourgeoisie, the exploiting and oppressing class.

Marx proposed that the elimination of discrimination among human beings could be achieved through a social system that abolished private property. The revolution that began in Paris in February 1848 spread to various countries, including Italy and Austria, but ultimately faced setbacks. Until Marx died in 1883, his longtime friend and unwavering supporter, Engels, stood by his side. However, they did not advocate for the complete cessation of the use of iron and steam engines, which were the foundations of capitalism.

4.2.2 Iron: An Accelerator Challenging the Culture of the United States

In the 19th century, the construction of railroads played a significant role in promoting exploration and encouraged pioneering and expansion toward the Western regions of the United States. The desire of pioneers to venture westward created a demand for transportation, which was further amplified by the US occupation of Texas after the war against Mexico in 1848 and the discovery of gold in California. The news of the gold discovery triggered a "Gold Rush" as miners, clad in jeans, flocked across the US.[70] Railroads emerged as a crucial means of meeting Americans' eagerness for gold and providing access to the West. The construction of railroads led to booming demand for labor, attracting a large influx of workers to the construction sites. The population growth in the West subsequently accelerated the territorial expansion of the region. Iron and steel production experienced rapid growth to meet the demand of the railroad industry, leading the US to surpass Britain in 1881 and become the world's largest producer of iron and steel.

Between 1850 and 1857, five railroad lines were established, connecting the Midwest and the eastern parts of the US, including routes that passed through the Appalachian Mountains. On May 10, 1869, the transcontinental railroad, linking the Atlantic and Pacific coasts, was completed, as depicted in Figure 4.22.[71] The introduction of the railroad significantly shortened the perilous six-month journey across the continent to a mere two weeks. In other words, the construction of the railroad effectively integrated the vast expanse of the US into a unified nation, fostering a

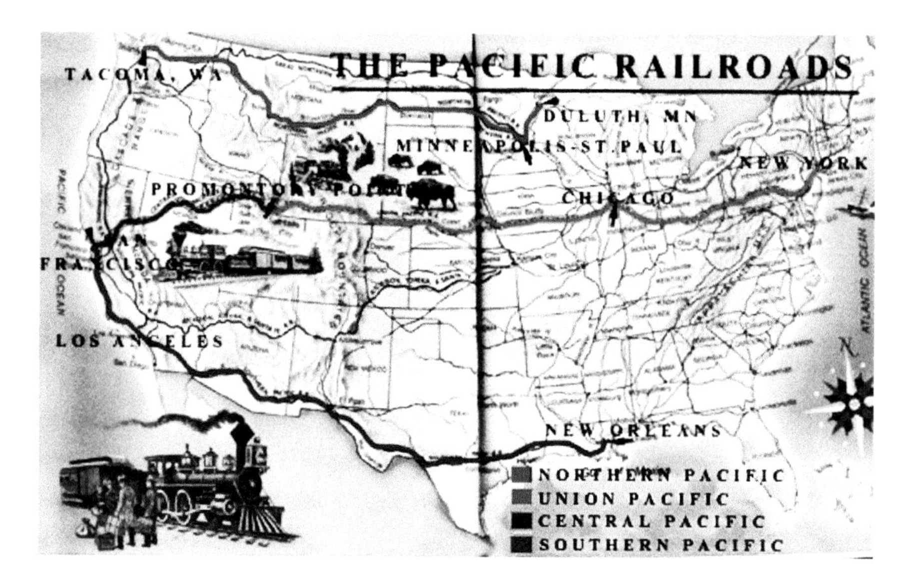

FIGURE 4.22 Intercontinental Railroads Built During Westward Expansion. (Courtesy of https://www.american-historama.org/1866-1881-reconstruction-era/transcontinental-railroad.htm)

shared way of life and cultural identity. The transportation revolution played a pivotal role in the emergence of the United States as a powerful and influential nation.

4.2.3 Iron Buddha: Mysterious Formative Beauty

An Iron Buddha refers to a Buddha statue that is cast in iron. In contrast to the gilt bronze Buddhas commonly found in temples, which are made of copper and coated with gold, Iron Buddhas possess a distinct sense of rawness, roughness, and power. Casting an Iron Buddha requires more energy, time, and money, and entails higher risks due to the higher melting point of iron compared to copper. This begs the question of why Iron Buddhas were created.

The use of iron as a material for casting Buddhist statues in Korean history was relatively limited, occurring mainly during the late Unified Silla and early Goryeo periods, from the 10th to the 11th century. Interestingly, the majority of Iron Buddhas discovered today are found in the central border area rather than in Gyeongju, the capital of Silla. While many Iron Buddhas were created during the Sui and Tang dynasties in China, most surviving examples date from the Song Dynasty onwards. In Japan, Iron Buddhas were crafted during the Kamakura period in the 13th century.

FIGURE 4.23 National Treasure No. 63 Iron Seated Vairocana Buddha at Dopiansa Temple in Cheorwon (left). Treasure No. 98 Iron Seated Buddha in Chungju (right). (Left, courtesy of http://www.heritage.go.kr/heri/cul/culSelectDetail.do?VdkVgwKey=11,00630000,32&pageNo=5_1_1_0#, right, courtesy of https://dh.aks.ac.kr/~heritage/wiki/index.php/%EC%B6%A9%EC%A3%BC_%EC%B2%A0%EC%A1%B0%EC%97%AC%EB%9E%98%EC%A2%8C%EC%83%81)

The popularity of Iron Buddhas emerged during the late Unified Silla period, characterized by instability in royal power in the capital and the growing influence of local powerful families.(72) The Iron Seated Buddhas depicted in Figure 4.23 were created during this era. Local influential families used Iron Buddhas as a metaphorical display of their strength, confidence, and defiance toward the central government and surrounding powers. Director Choi Wan-soo of the Kansong Art Museum stated that "the creation of Iron Buddhas at the time symbolically represented the accumulation of iron for the production of powerful weapons, and the appearance of the Iron Buddha embodied the image of the influential individuals who commissioned the statues."(73) Although more than a thousand years have passed since iron was first used on the Korean peninsula, the concept of using iron as a material to express the noble iconography of the Buddha emerged in the 9th century, 500 years after the spread of Buddhism in the region.

In China, the creation of Iron Buddhas was particularly prominent during the Tang Dynasty (713–743), and historical records mention the production of Iron Buddhas during the Song, Yuan, and Ming dynasties.[(74)] An inscription found in Shandong Province reveals that a statue of Zhang Yuk, an Iron Buddha, was made for the emperor in 563 AD, during the 2nd year of the Heqing reign of the Northern Qi dynasty in the Six Dynasties period. Chinese historical folklore, such as the "Taipingguangji," mentions the enshrinement of a 70-ft-tall Iron Buddha in the Great Buddha Temple during the Sui Huang year (581–600). Many regional records of Iron Buddha creation come from Shaanxi, an area rich in high-quality iron ore. In November 1994, during construction in Xian, Chengdu of Shaanxi Province, an underground well at the site of Jeongbeopsa Temple, built in 590 AD during the 10th year of the Sui Dynasty, yielded the discovery of the 'Iron seated statue of Maitreya Buddha' depicted in Figure 4.24 (left). This statue, estimated to have been made in the 8th century, is renowned

FIGURE 4.24 Seated statue of Maitreya Nyorai[(74)] in the Tang Dynasty (left) and Iron Buddha of the Kamakura Period in Japan[(75)] (right). (With permission from Professor Newspaper, http://www.kyosu.net/news/articlePrint.html?idxno=13156&page=quickViewArticleView)

for its excellent formative quality and distinctiveness from gilt-bronze Buddha statues due to the less conspicuous seams on its surface.[74]

In Japan, the creation of Iron Buddhas spanned from the Kamakura period (1185–1333) to the Muromachi period (1336–1573), and the tradition persisted until the Edo period (1603–1867).[75] The Kamakura period witnessed the highest prevalence of Iron Buddhas, which is attributed to the influence of Iron Buddha art from the Song Dynasty in China. Around 90% of the existing Iron Buddhas in Japan are concentrated in eastern regions. The earliest known Iron Buddha in Japan is the one depicted in Figure 4.24 (right), built in 1218.

4.2.4 Artistic Iron and Steel: Eiffel Tower and the Golden Gate Bridge

The most renowned artistic steel structure in the world is undoubtedly the Eiffel Tower, depicted in Figure 4.25. Not only is the Eiffel Tower an iconic symbol of France but it is also regarded as a global monument. Situated on the banks of the Seine River in Paris, the tower was conceived by Gustave Eiffel for the 1889 Exposition Universelle commemorating the 100th

FIGURE 4.25 Paris Eiffel Tower. (Courtesy of Wikimedia Commons, https://commons.wikimedia.org/wiki/File:Tour_Eiffel_Wikimedia_Commons_(cropped).jpg)

anniversary of the French Revolution. Before the construction of the Eiffel Tower, Eiffel had already made significant contributions to steel engineering. In 1884, he designed the Garabit viaduct in southern France, featuring a remarkable 165-m span, as well as the internal steel frame of the "Statue of Liberty" in New York and the variable dome of the Observatory in Nice, France. Due to his accomplishments, Eiffel earned the moniker "The Magician of Steel."[(76)] The French authorities received a total of 700 submissions for the fair's monumental structure, of which the Eiffel Tower was the only one they were satisfied with.

The construction of the Eiffel Tower involved the use of a concrete foundation supporting four iron pillars, upon which a steel structure was erected. Approximately 8,000 tons of materials were employed in its construction, with 7,300 tons of Puddled steel used for the tower's main body. The assembly process involved 18,038 wrought iron and steel pieces, which were joined together using around 2.5 million rivets. It took 300 workers 25 months to complete the tower, which stands at a magnificent height of 302 m. The Eiffel Tower remains a testament to the grandeur and engineering marvels achieved with steel.[(77)]

When the Eiffel Tower was initially constructed, it faced criticism from many who considered it a blemish on the harmonious and historic landscape of Paris. On February 14, 1887, Le Temps strongly condemned the Eiffel Tower, stating "To make our point, imagine for a moment a dizzying and ridiculous tower towering over Paris like a colossal black chimney, crushing Notre Dame, Tour Saint-Jacques, the Louvre, the Dome of Les Invalides, and the Arc de Triomphe. All our humiliated monuments will vanish in this horrifying vision. For twenty years, we will witness the expansion of the detestable shadow cast by this detestable column of bolted sheet metal."

Novelist Guy de Maupassant (1850–1893), who harbored a particular disdain for the Eiffel Tower, deliberately chose to have his lunch at a restaurant inside the tower itself. When asked why he dined there despite his hatred, he replied, "This is the only place in Paris where I cannot see the Eiffel Tower." Since 1985, the tower has been illuminated, providing a magnificent night view for visitors. After 130 years since its debut, the Eiffel Tower proudly stands in Paris as a cultural icon of the steel construction revolution.

The Golden Gate Bridge, depicted in Figure 4.26, was constructed to connect San Francisco and Marin County across the Golden Gate Strait, situated between San Francisco Bay and the Pacific Ocean. It was officially

FIGURE 4.26 Golden Gate Bridge in San Francisco. (Courtesy of Wikimedia Commons, https://en.wikipedia.org/wiki/File:GoldenGateBridge-001.jpg)

inaugurated on May 27, 1937, known as Pedestrian Day, and approximately 200,000 individuals crossed the bridge on its opening day. Soon after its completion, the Golden Gate Bridge swiftly became an iconic landmark of San Francisco, appearing in numerous movies and TV shows, including Superman, Godzilla, and Star Trek. In 1987, it was designated as a California Historic Landmark.

In 1996, the American Society of Civil Engineers recognized and announced the Seven Wonders of Modern Civil Architecture, with the Golden Gate Bridge being one of the selected structures. The construction of this steel suspension bridge, spanning 2,825 m in length and 27 m in width, commenced in 1933. Plans for a bridge over the Golden Gate Strait had been contemplated since 1872, but suitable forms, such as steel cable suspension bridges, had not yet been adequately developed.

American civil engineer Joseph Strauss, who had already designed 400 bridges in the early 1920s, proposed a steel suspension bridge with an extensive span, referring to the distance between piers. The Golden Gate Bridge was designed with a span of 1,280 m, more than twice the length of the longest bridge at that time. Many architects and engineers raised concerns about its safety, deeming it a hazardous venture. However, after extensive discussions, Strauss's vision was realized, and the grand construction project commenced on January 5, 1933.

Strauss encountered numerous challenges throughout the construction period. The swift tides, frequent storms, and dense fog in the area hindered the smooth progress of the project. A cargo ship collided with one of the support bridges during construction, resulting in significant damage. Moreover, since the location falls within the San Andreas fault zone, earthquakes were a common occurrence, necessitating careful attention to earthquake-resistant design. In response, Strauss implemented the most stringent safety measures ever seen in the history of bridge construction.

To safeguard the lives of the construction workers, Strauss installed a safety net beneath the bridge, preventing them from being swept away by the wind. However, on February 17, 1937, just three months before the bridge's opening, twelve workers tragically fell through the safety net due to the collapse of the bridge scaffolding, resulting in the loss of ten lives. Despite these setbacks, the Golden Gate Bridge, a suspension bridge with two steel cables hanging from a tower measuring 227 m in height, was ultimately opened in 1937.

The wire used to support the bridge spanned a length of 2,332 m and had a diameter of 92 cm. The cable consisted of 27,572 very thin wires twisted together, totaling a length of 128,748 km. The cables used solely for the Golden Gate Bridge weighed 24,500 tons.[(78)] Through the construction of the Golden Gate Bridge, we come to appreciate that steel is not merely a strong material but also capable of creating structurally and aesthetically beautiful bridges. Since the completion of the Golden Gate Bridge, numerous suspension bridges with even longer spans have been constructed worldwide, many of which have become iconic landmarks in their respective regions.

4.2.5 Public Art, Iron Sculpture

Iron can be transformed into exquisite sculptures that bring joy to the public through appreciation. In recent years, the concept of "Public Art" has gained popularity, leading to the installation of sculptures in parks across cities worldwide, including Korea, with iron being a widely used material. The term "public art" was coined by British author John Willett (1917–2002) in his 1967 book "Art in a City."[(79)] Willett introduced public art as a novel concept that engages the emotions of the general public, challenging the notion that only a few experts such as art directors and critics can represent the public's sense of beauty. This idea was further realized through two programs initiated by the US government in the late 1960s: the "Special Equity Investment for Art" program and the "Art in Public Place" program by the National Endowment for the Arts. The equity investment program for art is currently implemented in various cities and counties in the US, where a certain percentage, typically 1%, of the construction budget is allocated for art when a new public building is constructed.(80)

FIGURE 4.27 (Top) City of Chicago's "Cloud Gate" made of stainless steel.(79) (Bottom, left) Fist of a Champion, Detroit, MI. (Bottom, center) Seoul POSCO Center "Flower Blooming Structure," (Bottom, right) "Hammering man" in front of Heungkuk Life Insurance Building in Seoul, South Korea. (top, courtesy of https://unsplash.com/ko/%EC%82%AC%EC%A7%84/AlEPagnXhXw, bottom left, courtesy of https://en.m.wikipedia.org/wiki/File:Monument_to_Joe_Louis--.jpg)

Figure 4.27 showcases examples of iron sculptures transformed into public art. Two of the works are located in Detroit and Chicago, United States, while the other two are situated in Seoul, Korea. These sculptures not only serve as artistic expressions but also merge harmoniously with their respective settings. For instance, Jonathan Borofsky's "Hammering Man" and Henry Moore's multiple works have been produced and installed in various locations. The "Fist Sculpture" is located in downtown Detroit, the "Cloud Gate" graces a park in Chicago, and the "Hammering Man" stands proudly in front of the Heungkuk Life Insurance Building in Gwanghwamun, Seoul.

REFERENCES

1. "Control of fire by early humans," last edited March 9, 2024, https://en.wikipedia.org/wiki/Control_of_fire_by_early_humans
2. "Fertile crescent," last edited March 28 , 2024, https://en.wikipedia.org/wiki/Fertile_Crescent
3. Alan W. Cramb, "A short history of metals," https://vietnamclassical.files.wordpress.com/2011/10/history-of-metals1.pdf
4. D. J. C. Taylor, D. C. Page, and P. Geldenhuys, "Iron and steel in South Africa," *J. S. Afr. Inst. Min. Metall.*, Vol. 88, No. 3, March 1988. pp. 73–95, https://www.saimm.co.za/Journal/v088n03p073.pdf
5. Augustin F. C. Holl, "Origins of African metallurgy, Oxford research encyclopedia of anthropology subject: archaeology", doi:10.1093/acrefore/9780190854584.013.63
6. Stanley B. Alpern, "Did they or didn't they invent it? Iron in sub-Saharan Africa", *History in Africa*, Vol. 32, 2005, pp. 41–94, https://www.researchgate.net/publication/236832306_Did_They_or_Didn't_They_Invent_It_Iron_in_Sub-Saharan_Africa
7. Aslı Özyar, "Anatolia from 2000 to 550 BCE", Chapter 3.10, Cambridge University Press, pp. 1545–1570, http://dx.doi.org/10.1017/CHO9781139017831.095
8. Violetta Cordani, "The development of the Hittite iron industry. A reappraisal of the written sources", *Die Welt des Orients*, Vol. 46, No. 2, 2016, pp. 162–176, http://www.jstor.com/stable/24887872
9. Thomas R. Martin, "Ancient Greece from Prehistoric to Hellenistic Times", 2nd ed., New Haven and London: Yale University Press, http://elibrary.bsu.edu.az/files/books_163/N_23.pdf
10. T. Rehren, T. Belgya, A. Jambon, G. Káli, Z. Kasztovszky, Z. Kis, I. Kovács, B. Maróti, Marcos M. Torres, G. Miniaci, Vincent C. Pigott, M. Radivojevic, L. Rosta, L. Szentmiklósi, and Zoltán S. Nagy, *J. Archaeol. Sci.*, Vol. 40, 2013, pp. 4785–4792, https://www.sciencedirect.com/science/article/pii/S0305440313002057

11. N. L. Erb-Satullo, "The innovation and adoption of iron in the ancient near East", *J. Archaeol Res.*, Vol. 27, 2019, pp. 557–607, https://doi.org/10.1007/s10814-019-09129-6
12. "Axial age", last edited March 14, 2024, https://en.wikipedia.org/wiki/Axial_Age
13. F. B. Jevons, "Iron in homer", *J. Hell. Stud.*, Vol. 13, 1892–1893, pp. 25–31, https://www.jstor.org/stable/623888#metadata_info_tab_contents
14. Henry Gary, "The Histories of Herodotus", New York D. Appleton and Company, 1904, https://ia800301.us.archive.org/15/items/historiesofherod00hero/historiesofherod00hero.pdf
15. "The Etruscans on Elba", https://www.elbaworld.com/en/elba-history/p-378-etruscans-on-elba.html
16. "Aristotle, minor works", *Chalybes*, Vol. 832, p. 23, Loeb Classical Library, https://www.loebclassics.com/view/LCL307/1936/pb_LCL307.509.xml
17. "What was Greek philosophy before Socrates?", https://www.wondrium-daily.com/greek-philosophy-before-socrates/
18. C. George Boeree, "The ancient Greeks, part two: Socrates, Plato, and Aristotle", http://webspace.ship.edu/cgboer/athenians.html
19. E. A. Ginzel, "Steel in ancient Greece and Rome", 1995, http://dtrinkle.matse.illinois.edu/MatSE584/articles/steel_greece_rome/steel_in_ancient_greece_an.html
20. Philippe Beaujard, "From three possible iron-age world-systems to a single Afro-Eurasian world-system", *J. World Hist.*, Vol. 21, No. 1, 2010, https://www.jstor.org/stable/20752924
21. Janet Lang, "Paper for special issue on 'aspects of ancient metallurgy' Roman iron and steel: A review", *Mater. Manuf. Process.*, 2017, doi:10.1080/10426914.2017.1279326
22. F. Becker et al, "The furnace and the goat—A spatio-temporal model of the fuelwood requirement for iron metallurgy on Elba Island, 4th century BCE to 2nd century CE", *PLoS One*, Vol. 15, No. 11, 2020, p. e0241133, https://doi.org/10.1371/journal.pone.0241133
23. Adrian Tan, "A computer-generated model of the construction of the roman colosseum" (Electronic Thesis or Dissertation), 2012, https://www.semanticscholar.org/paper/A-Computer-Generated-Model-of-the-Construction-of-Tan/38e8ade1621ece391e57c97dec762c8947fabc02#extracted
24. Janet Lang, "Study of the Metallography of Some Roman Swords", *Britannia*, Vol. 19, 1988, pp. 199–216, https://www.jstor.org/stable/526199
25. Raoul McLaughlin, "Rome and the distant east: Trade routes to the ancient lands of Arabia, India and China", *Continuum*, July 8, 2010, https://www.amazon.com/Rome-Distant-East-Routes-ancient/dp/1847252354
26. "Han dynasty iron production", Jul 20, 2011, https://historum.com/t/han-dynasty-iron-production.28104/
27. Kevin Greene, "The Archaeology of the Roman Economy", University of California Press, December 1990, https://www.ucpress.edu/book/9780520074019/the-archaeology-of-the-roman-economy#about-book

28. K. Vaish et. al., "Historical perspective of iron in ancient India", *J. Met. Mater. SC.*, Vol. 42, No. 1, 2000, pp. 65–74, https://eprints.nmlindia.org/1471/2/65-74.PDF
29. "Buddhism and Hinduism", last edited March 25, 2024, https://en.wikipedia.org/wiki/Buddhism_and_Hinduism
30. "Wootz Damascus steel: The mysterious metal that was used in deadly blades", https://www.ancient-origins.net/artifacts-ancient-technology/wootz-steel-damascus-blades-0010148
31. John Noble Wilford, "Ancient smelter used wind to make high-grade steel", February 6, 1996, https://www.nytimes.com/1996/02/06/science/ancient-smelter-used-wind-to-make-high-grade-steel.html
32. Donald B. Wagner, "The earliest use of iron in China", Published in *Metals in Antiquity*, edited by Suzanne M. M. Young, A. Mark Pollard, Paul Budd, and Robert A. Ixer (BAR international series, 792), Oxford: Archaeopress, 1999, pp. 1–9, http://donwagner.dk/EARFE/EARFE.html
33. K. Chen, J. Mei, P. Wang, L. Wang, Y. Wang, and Y. Liu, "Archaeometallurgical studies in China: Some recent developments and challenging issues", *J. Archaeol. Sci.*, Vol. 56, 2015, pp. 221–232, https://doi.org/10.1016/j.jas.2015.02.026
34. "Iron making history in China", http://en.chinaculture.org/library/2008-02/01/content_26524.htm
35. Wei Qian and Xing Huang, "Invention of cast iron smelting in early China: Archaeological survey and numerical simulation", *Adv. Archaeomater.*, Vol. 2, No. 1, 2021, pp. 4–14, ISSN 2667-1360, https://doi.org/10.1016/j.aia.2021.04.001
36. Donald B. Wagner, "Cast Iron in China and Europe", Symposium on Cast Iron in Ancient China, Beijing, July 20, 2009, http://donwagner.dk/cice/cice.html
37. Donald B. Wagner, "Technology as seen through the case of ferrous metallurgy in Han China", http://donwagner.dk/EncIt/EncIt.html
38. Yaxiong Liu, "Iron production in the state of Qin during the Warring States period", Thesis submitted to University College London for the Degree of Doctor of Philosophy UCL Institute of Archaeology, February 9, 2021, https://discovery.ucl.ac.uk/id/eprint/10130408/1/Liu_10130408_Thesis_revised.pdf
39. "Economy of the Han dynasty", last edited March 5, 2024, https://en.wikipedia.org/wiki/Economy_of_the_Han_dynasty
40. "Analects", last edited March 26, 2024, https://en.wikipedia.org/wiki/Analects
41. "Confucianism", https://education.nationalgeographic.org/resource/confucianism
42. "Three kingdoms", Vol 30 (in Chinese), https://zh.wikisource.org/wiki/%E4%B8%89%E5%9C%8B%E5%BF%97/%E5%8D%B730
43. "History of the North", Vol. 94 (in Chinese), https://zh.wikisource.org/wiki/%E5%8C%97%E5%8F%B2/%E5%8D%B7094

44. S. T. Kim, "A study on Goguryeo weapons excavated in South Korea" (in Korean), https://memory.library.kr/files/original/71a99efe6ba02417b710aea2ee98e6ec.pdf
45. "Excavation of 9 additional smelting furnaces from the 3rd and 4th centuries in Chilgeum-dong, Chungju", *Press Release, National Jungwon Cultural Heritage Research Institute*, November 21, 2018 (in Korean), https://nrich.go.kr/jungwon/boardView.do?menuIdx=726&bbscd=33&bbs_idx=41022
46. "Chiljido" (in Korean), last edited March 20, 2024, https://ko.wikipedia.org/wiki/%EC%B9%A0%EC%A7%80%EB%8F%84
47. "Developmental aspects of Baekje iron culture and iron manufacturing technology", *The Korean Iron Culture Research Association*, November 23, 2013 (in Korean), https://opengov.seoul.go.kr/research/6413981
48. B. C. Woo, "Ironware culture of ancient Ulsan", Ulsan Soeburi Festival Academic Symposium, May 4 2013 (in Korean), http://www.soeburi.org/media/2013_%E1%84%92%E1%85%A1%E1%86%A8%E1%84%89%E1%85%AE%E1%86%AF%E1%84%89%E1%85%B5%E1%86%B7%E1%84%91%E1%85%A9%E1%84%8C%E1%85%B5%E1%84%8B%E1%85%A5%E1%86%B7%E1%84%8B%E1%85%AE%E1%86%AF%E1%84%89%E1%85%A1%E1%86%AB%E1%84%8B%E1%85%B4%E1%84%89%E1%85%AC%E1%84%87%E1%85%AE%E1%84%85%E1%85%B5%E1%84%86%E1%85%AE%E1%86%AB%E1%84%92%E1%85%AA.pdf
49. "Dalcheon iron complex, the main axis of Korea's iron production", Gyeongnam, Busan, Ulsan Cultural Heritage Story Tour, Cultural Heritage Administration Book, pp 359–363, in Korean, https://www.cha.go.kr/cop/bbs/selectBoardArticle.do?nttId=63999&bbsId=BBSMSTR_1222&mn=NS_03_08_03&ccbaCpno=2332600400000®ionGbn=26&pageUnit=10&pageIndex=1&rnum=1)
50. M. Ozawa, "Irons in Japan and the Korean Peninsula based on metallurgy", *National Museum of Japanese History Research Report*, Vol. 110, February 2004 (in Japanese), https://www.google.co.kr/url?sa=t&rct=j&q=&esrc=s&source=web&cd=&ved=2ahUKEwiV5NPUtq38AhXG1GEKHTWrA9EQFnoECAoQAQ&url=https%3A%2F%2Frekihaku.repo.nii.ac.jp%2Findex.php%3Faction%3Dpages_view_main%26active_action%3Drepository_action_common_download%26item_id%3D1204%26item_no%3D1%26attribute_id%3D22%26file_no%3D1%26page_id%3D13%26block_id%3D41&usg=AOvVaw11QLUFFiz9MMDAIIjH1MV9
51. "The power of a great nation, cast iron, *YTN Science*", 2016 (in Korean), https://m.science.ytn.co.kr/program/view.php?mcd=0033&key=201608161039193361
52. "Three treasures in Shilla" (in Korean), https://namu.wiki/w/%EC%8B%A0%EB%9D%BC%EC%82%BC%EB%B3%B4
53. "Yayoi period", last edited March 23, 2024, https://en.wikipedia.org/wiki/Yayoi_period

54. Tatsuo Inoue, "Tatara and the Japanese sword: The science and technology", *Acta Mechanica*, Vol. 214, 2010, pp. 17–30, https://www.semanticscholar.org/paper/Tatara-and-the-Japanese-sword%3A-the-science-and-Inoue/b02cb7175aef918d7fc291706f6113ddd8b8fd62
55. Leon Kapp, Hiroko Kapp, and Yoshindo Yoshihara, "The Craft of the Japanese Sword", Kodansha International Ltd., 1987, pp. 61–66, https://pdf-coffee.com/the-craft-of-the-japanese-sword-pdf-free.html
56. "Toyotomi Hideyoshi's Sword hunting" (in Japanese), https://www.meihaku.jp/sword-basic/hideyoshi-katanagari/
57. "Swords as gifts", Samurai Archives Japanese History Forum, Official forum of the Samurai Archives Japanese History page, https://www.tapatalk.com/groups/japanese_history/swords-as-gifts-t669.html
58. "Venetian navy", last edited March 1, 2024, https://en.wikipedia.org/wiki/Venetian_navy
59. "Venetian arsenal", last edited February 10, 2024, https://en.wikipedia.org/wiki/Venetian_Arsenal
60. "House of Medici", last edited March 29, 2024, https://en.wikipedia.org/wiki/House_of_Medici
61. "Ninety-five theses", last edited March 14, 2024, https://en.wikipedia.org/wiki/Ninety-five_Theses
62. "Johannes Gutenberg: The first printing press and publishing", https://malevus.com/johannes-gutenberg/
63. "The Gutenberg Press: An invention that changed the world", https://www.museumofthebible.org/book-minute/the-gutenberg-press-an-invention-that-changed-the-world
64. "Age of Enlightenment", last edited March 29, 2024, https://en.wikipedia.org/wiki/Age_of_Enlightenment
65. "Cries of pain: The word 'capitalism'", https://www.cambridge.org/core/books/abs/information-nexus/cries-of-pain-the-word-capitalism/67D26C2D04F0B410AD317F777E53CE41
66. "Adam Smith and 'The wealth of nations'", https://www.investopedia.com/updates/adam-smith-wealth-of-nations/
67. Mark Brayshay, "Capitalism and the division of labor", Audrey Kobayashi (Editor), *International Encyclopedia of Human Geography*, 2nd ed., Elsevier Science & Technology, 2020, pp. 23–41.
68. John P. Rafferty, "The rise of the machines: Pros and cons of the industrial revolution", https://www.britannica.com/story/the-rise-of-the-machines-pros-and-cons-of-the-industrial-revolution
69. "Partnership with Marx", https://www.britannica.com/biography/Friedrich-Engels/Partnership-with-Marx
70. "California gold rush", Editors updated August 10, 2022, original April 6, 2010, https://www.history.com/topics/westward-expansion/gold-rush-of-1849
71. "First transcontinental railroad", last edited March 30, 2024, https://en.wikipedia.org/wiki/First_transcontinental_railroad
72. H. C. Yang, "Cast-iron Buddha", by The National Museum of Korea, https://smarthistory.org/cast-iron-buddha-goryeo/

73. W. S. Choi, “Choi Wan-su’s view of our culture”, Vol. 26 (in Korean), http://1004eyes.com/bbs/view.php?id=eng_history&page=2&sn1=&divpage=1&sn=off&ss=on&sc=on&select_arrange=headnum&desc=asc&no=151
74. K. M. Kwon, “The new examples of the iron Buddha’s made in the tang dynasty China”, *Art Data*, 2010, 146–162 (in Korean), https://www.museum.go.kr/site/main/archive/periodical/archive_6208
75. S. E Choi, “Comparison with Chinese and Japanese Iron Buddha’s”, Deoksung Women’s University, 2007 (in Korean), http://www.kyosu.net/news/articlePrint.html?idxno=13156&page=quickViewArticleView
76. “Gustave Eiffel”, last edited February 23, 2024, https://en.wikipedia.org/wiki/Gustave_Eiffel
77. “Eiffel tower”, last edited March 9, 2024, https://en.wikipedia.org/wiki/Eiffel_Tower
78. “Golden gate bridge”, last edited March 29, 2024, https://en.wikipedia.org/wiki/Golden_Gate_Bridge
79. John Willett, “Art in a City”, Liverpool University Press, 2007, https://books.google.co.kr/books/about/Art_in_a_City.html?id=btzpAAAAMAAJ&redir_esc=y
80. “20+ Places to find spectacular public art around the U.S.”, https://mymodernmet.com/public-art-in-the-us/

CHAPTER 5

Iron – Key Partner of the Industrial Revolution

5.1 STEEL LEADING THE INDUSTRIAL REVOLUTION

The knowledge of iron smelting reached the British Isles through the Celts in the 8th century BC. The Celts gained an advantage in battles due to their formidable iron weapons. As the Celts spread iron civilization, European farmers had more time for other activities alongside agriculture. This newfound free time was utilized for salt production, clothing manufacturing, and crafting extravagant jewelry. Many of these goods were exchanged through long-distance trade.

Today, iron plays a vital role as a material for machines operating within factories, rather than merely serving as a construction material for factories and buildings. The British steel industry, with its abundant iron reserves, initially relied on bloomery furnaces. However, with the development of coke and blast furnace (BF) ironmaking, Britain was able to rise to industrial power earlier than neighboring countries. Innovations were made in the transportation of iron ore, coal, and heavy iron products required by the ironworks where BFs operated. The commercialization of the Bessemer steelmaking process in the late 1800s facilitated the rapid conversion of ferrous products into steel. This led to the production of

DOI: 10.1201/9781003419259-5

strong, abrasion-resistant steel rails, enabling the era of mass transportation with the practical utilization of steam engines.

Iron played a pivotal role in the Industrial Revolution that originated in England as it was utilized in the production of steam engines, railroads, trains, and machinery. The steam engine made it possible to efficiently operate heavy iron-based textile equipment. Additionally, the invention of steam-powered locomotives, like the one depicted in Figure 5.1, revolutionized mass transportation. Iron ships, powered by steam engines, facilitated long-distance travel. Iron formed the foundation of all the necessities required for the Industrial Revolution, leading to corresponding innovations in iron and steelmaking technology.

The widespread use of steam engines, particularly in power machines such as pumps, further increased the demand for iron. Steam engine pumps proved to be excellent for drainage, alleviating a major issue in mines at the time. Railroads expanded extensively, connecting steel mills, mines, and residential areas, ushering in an era of mass transportation. The steam engine enabled the efficient introduction of large amounts of air into BFs, enhancing iron productivity. The symbiotic relationship between iron and steam engines fostered a virtuous cycle across all industrial fields,

FIGURE 5.1 A steam locomotive equipped with a steam engine became the driving force of the Industrial Revolution. (Courtesy of https://commons.wikimedia.org/wiki/File:Flying_Scotsman_in_Doncaster.JPG)

driving the demand and fueling the flames of the Industrial Revolution, which spread across the world.

Railroads stimulated the economy in two significant ways. First, the introduction of efficient transportation reduced the cost of transporting goods, leading to a decrease in product prices and subsequently an increase in demand. The rising demand resulted in greater utilization of factories and expansion of facilities, which, in turn, increased the demand for machinery such as steam engines and carbon fuel to power them.

Second, the demand for iron and steel skyrocketed. These materials were essential for constructing railroads, locomotives, drainage pumps in mines, steam engines in factories, and steamboats. The widespread use of steam engines would not have been as successful without the technological advancements in the iron industry. All these interconnected developments played a crucial role in fueling the Industrial Revolution. Innovations in one field propelled progress in other fields.

The early iron industry was established and operated in forested areas where charcoal, the primary fuel source, was readily available. It was more cost-effective to transport a small volume of iron or steel to the production plant than to move a larger volume of charcoal. However, due to the rapid depletion of forests caused by the mass production of iron, the search for alternative fuels became urgent. As discussed in Chapter 3, in 1709, British iron master Abraham Darby I (1678–1717) developed the use of coke in ironmaking to solve this problem.

This innovation led to a significant increase in the number of coke ovens, from 61 in 1717 to 236 in 1805. Ironworks gradually shifted their locations to areas with a steady supply of coal, and approximately 75% of ironworks were situated near coal fields. This shift, coupled with the advancement of steelmaking methods discussed in Section 3.3, propelled Britain's steel industry to remarkable growth by 1800. Britain emerged as a dominant power leading the Industrial Revolution, accounting for half of the world's iron production. Meanwhile, another significant innovation actively developed during this time was the steelmaking method, which resulted in a substantial improvement in the quality of steel.

5.2 DEVELOPMENT OF THE IRON STEAM ENGINE

The iron steam engine served as the fundamental catalyst for the Industrial Revolution in 18th-century England. The widespread adoption of the steam engine was made possible by the production of high-quality iron and steel at a low cost. British crude steel production witnessed a remarkable increase from 573,301 tons in 1873 to 4.9 million tons in 1900.[1] This exponential growth established Britain as a leading global power in the spinning industry during the Industrial Revolution.

5.2.1 The Precursor to the Steam Engine

The steam engine, an iconic symbol of Britain's Industrial Revolution, was a complex invention that incorporated various innovations. In 1644, Evangelista Torricelli (1608–1647), a follower of Galileo Galilei (1564–1642), made a significant contribution by measuring atmospheric pressure in 1644 using mercury, as depicted in Figure 5.2. Torricelli's experiment involved filling a glass tube, approximately 1 m in height and closed at one end, with mercury. When this tube was inverted and placed in a container of mercury, the mercury in the glass tube settled at a height of approximately 76 cm from the surface. This equilibrium occurred due to the balance between the atmospheric pressure exerted on the mercury in the container and the force exerted by the mercury in the tube.[2]

German physicist and politician Otto von Guericke (1602–1686) conducted an experiment that demonstrated the significant difference between vacuum and atmospheric pressure. In 1654, he performed the "Magdeburg hemisphere experiment" using two hemispheres with a diameter of 50 cm,

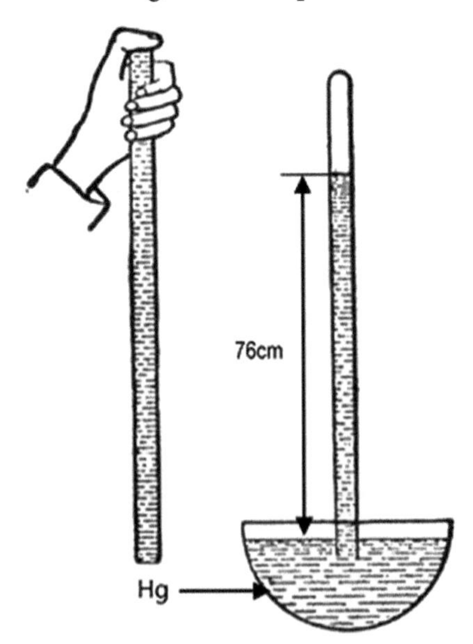

FIGURE 5.2 Torricelli mercury test.[2] (Courtesy of https://en.wikipedia.org/wiki/Evangelista_Torricelli#/media/File:NSRW_Torricelli's_experiment.jpg)

FIGURE 5.3 Magdeburg hemisphere experiment.[3] (Courtesy of https://commons.wikimedia.org/wiki/File:Magdeburg.jpg)

as depicted in Figure 5.3. Guericke constructed two hemispheres made of oil-soaked leather and placed them together with the rings in contact. The air between the hemispheres was then evacuated to create a vacuum. The resulting vacuum hemisphere exerted such a tremendous force that it was extremely challenging to separate the hemispheres even when eight horses on each side pulled with all their might.[3]

Denis Papin (1647–1712), a French physicist, proposed an early steam engine that utilized the force of steam and the resulting pressure difference caused by its condensation. This device operated on the principle that as water vapor flowed over the piston, the pressure would push the piston downward. Conversely, when the vapor above the piston cooled and condensed due to contact with water beneath the piston, the pressure would decrease, causing the piston to rise.[4] Papin's invention was essentially a pumping device capable of lifting water. Although his device did not find practical use, it laid the foundation for subsequent improvements by Savery, Newcomen, and Watt.

5.2.2 From the Savery Engine to the Newcomen Engine

Thomas Savery (1650–1715), an English engineer and inventor with a particular interest in draining water from coal mines, invented a water pump based on the principles of the steam engine in 1693 and patented it in 1698. Instead of introducing water into the cylinder, Savery designed a pumping machine where the boiler produced steam that was then directed to the cylinder and subsequently cooled, as illustrated in Figure 5.4.[5] Around 1700, Savery expanded his research to explore applications beyond coal mine drainage, including water supply and water turbine power.

Savery envisioned that if his engine were used in three stages, it could successfully lift water from a depth of up to 70 m. However, in practical trials conducted in several coal mines, the engine proved incapable of lifting water even at depths of 20 to 30 m. This was primarily due to the increased pressure and heat of the steam, which resulted in container

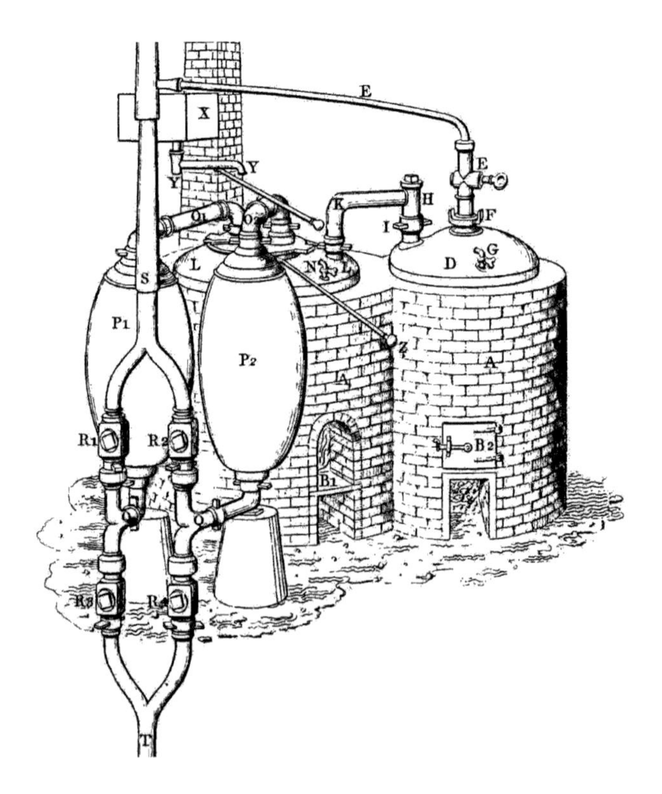

FIGURE 5.4 Savery's steam engine.[5] (Courtesy of https://commons.wikimedia.org/wiki/File:Savery's_Steam-Engine_from_Farey-Plate01cut3.jpg)

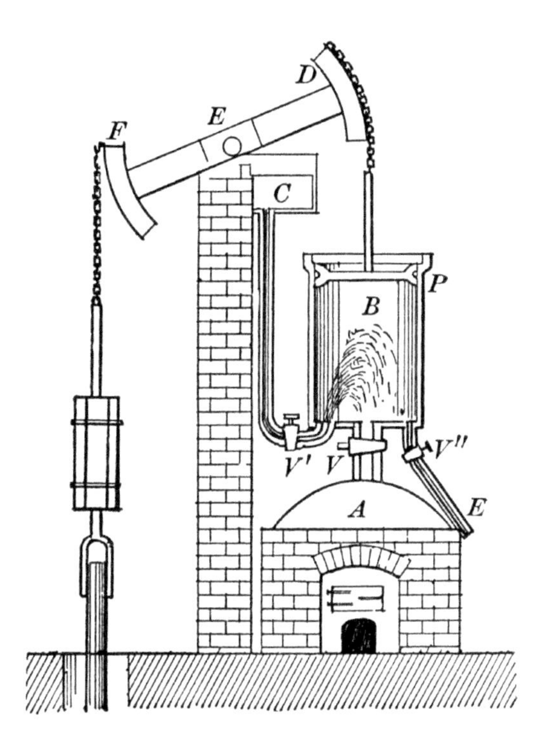

FIGURE 5.5 Newcomen's steam engine.[6] (Courtesy of https://en.m.wikipedia.org/wiki/File:Newcomen6325.png)

explosions and melted welds. Nevertheless, a few years later, Savery, along with Thomas Newcomen, developed a new engine that addressed these issues, known as the Newcomen engine.[6]

The Newcomen engine, depicted in Figure 5.5, was introduced in 1712 and became the most efficient steam pump of its time. Towards the end of the 17th century, the scarcity of wood throughout the country led to an increased demand for coal as an alternative fuel source. Consequently, coal mining operations were forced to delve deeper underground. However, these deep tunnels were often inundated with water. The pump made of the New Common engine effectively drained these water-filled tunnels. Additionally, the engine itself was found to be fueled using coal dust that had been treated as waste from the coal mines. This technological advancement played a pivotal role in facilitating the rapid transition from wood to coal as the primary fuel for British industry.

From the initial implementation of Thomas Newcomen's steam engine at Dudley Castle Coal Mine in 1712 to the advent of the internal combustion engine in the early 20th century, the steam engine served as the primary driving force behind the Industrial Revolution. Newcomen's steam engine operated based on an atmospheric pressure method. High-pressure steam was introduced into the lower part of the cylinder, causing the piston to rise. As the steam cooled and its pressure reduced, the resulting negative pressure moved the piston.

5.3 WATT'S STEAM ENGINE: THE CORE OF THE INDUSTRIAL REVOLUTION

James Watt (1736–1819) played a crucial role in improving the Newcomen steam engine. While working in a repair shop at the University of Glasgow, Watt was tasked with repairing a Newcomen engine. However, even after the repairs, the engine still had low operating efficiency. The Newcomen engine operated on the principle of atmospheric pressure, with the piston reciprocating based on the compression and expansion of water vapor in the cylinder. Additionally, the entire cylinder of the Newcomen engine was cooled, resulting in significant heat loss and high coal consumption.

On January 5, 1769, Watt patented his invention titled "Method of lessening the consumption of steam and fuel in a fire engine."(7) He devised a method in which the steam was cooled in a separate condenser connected to the cylinder, allowing the piston to move based on steam pressure. This innovation conserved cylinder heat and significantly increased efficiency, reducing coal consumption to less than a quarter of that required by the Newcomen engine. Furthermore, Watt developed a double-acting steam engine in which force acted on both the upward and downward movement of the piston.

Watt commenced the production of engines in 1775 through a partnership with Matthew Boulton (1728–1809), a manufacturer and engineer based in Birmingham. During this period, practical applications of ironmaking in BFs, as well as the steelmaking techniques described by Huntsman and Cort in Chapter 3, were in full swing. Consequently, affordable and high-quality malleable cast iron became the preferred material for steam engines, playing a pivotal role in the manufacturing of highly efficient steam engines.

In a letter from Boulton to Watt in 1781, he expressed the enthusiasm of the people in London, Manchester, and Birmingham for works operated by steam engines.(8) This inspired Watt to equip his engines with speed-controlled wheels, driven by a beam through a set of "sun-and-planet gears," enabling rotational motion.[9] Inventor Richard Arkwright (1732–1792) later adapted and improved this system to suit spinning machines, leading to its rapid adoption in textile factories.[10] Before this innovation, factories relied heavily on water or animal power to operate their heavy machinery.

The development of steam engines and the expansion of their applications continued to progress. Watt made significant advancements by introducing a centrifugal governor to regulate the engine's speed and a parallel motion device to maintain the piston's alignment within the cylinder. These innovations led to the production of over 500 steam engines, as depicted in Figure 5.6, between 1775 and 1800.[11] Watt and Boulton transformed the steam engine from a simple pump into a high-performance and versatile prime mover that could be applied to a wide range of industrial processes. The improvements made to the steam engine by Boulton and Watt enabled the development of various machines, including factory machinery, steam locomotives, steamboats, and automobiles. As a result, the transportation of people and goods became significantly faster than before. Watt often predicted that "If the steam engine works properly, there will be great changes in the world," and as he anticipated, the steam engine brought about a dramatic transformation in the world.

FIGURE 5.6 James Watt's steam engine. (Courtesy of Wikipedia, https://en.m.wikipedia.org/wiki/File:20070616_Dampfmaschine.jpg)

The steam engine continued to undergo advancements even after 1800. Richard Trevithick, an engineer from Cornwall, played a significant role in the commercialization of steam engines for locomotives that could safely operate at higher pressures compared to Watt's engine. In 1804, Trevithick's first steam engine was utilized on the Pen-y-Darren railway in south Wales, and it became known as the "Cornish engine." This engine operated based on the Cornish cycle,[12] where pressurized steam was supplied to the top part of the cylinder to lower the piston with the help of counterweights, and pressurized steam was supplied to the bottom part of the cylinder to raise the piston.

The steam engine evolved into a machine capable of performing a wide range of functions by actively responding to customer demands. By the mid-19th century, steam engines dominated both land and sea transportation and became indispensable components of factories. These steam engines were primarily made of iron and steel. As the Industrial Revolution expanded, Britain became a global leader in iron and steel production. The development of converter steelmaking by Bessemer and Thomas in the mid-19th century significantly improved the quality of steel, further enhancing the performance and efficiency of steam engines while also reducing their costs. Iron played a crucial role in supporting the Industrial Revolution, and as the revolution progressed, the demand for iron increased, creating a mutually reinforcing relationship between iron production and the Industrial Revolution.

5.4 STEEL REVOLUTIONIZED TRANSPORTATION

5.4.1 Evolution of Wheels for Automobiles

As a means of transportation for goods and people, the wheel is considered one of humanity's greatest inventions. The earliest known wheel as an artifact was discovered in Mesopotamian ruins dating back to around 3500 BC. The use of wheels for transportation purposes can be traced back to the employment of chariots in Mesopotamia around 3200 BC. Even during the Paleolithic period (750,000 to 15,000 years ago), wheels were utilized by our ancestors, who used logs to move large stones.[13] However, it took around 1,500 years for the concept of the wheel to evolve. Before that, wheels were typically disc-shaped and crafted by baking soil or carving wood. The introduction of spoked wheels for chariots occurred in the Eurasian Steppe around 2000 BC.[14]

The first iron-rimmed wheel is believed to have originated from a Hittite chariot around 1300 BC, although no physical remains of it have been found. The wheels of Celtic chariots around 1000 BC were equipped with iron frames.[15] This design remained largely unchanged for nearly 2,000 years, with incremental increases in the proportion of iron used in the rim and axle. However, in 1802, G. F. Bauer patented his "elastic wheels," which replaced spokes with leather straps, resulting in a significant change in the wheel's appearance.[16] In 1845, Robert William Thomson (1822–1873) invented the vulcanized rubber pneumatic (inflatable) tire, which he patented. This innovation greatly improved the ride comfort of wagons when applied to their wheels.[17]

Karl Benz (1844–1929), a German engineer, invented a gas engine in 1878 and subsequently built the first three-wheeled vehicle, known as the Motorwagen (depicted in Figure 5.7), using a gasoline engine in 1885.[18] He operated the Mercedes-Benz factory in Mannheim, Baden-Württemberg, Germany, and later established the Daimler-Benz factory in 1926 through a joint venture with Daimler. Benz's initial three-wheeled car featured wire wheels similar to those found on bicycles, with hard rubber wrapped around the rims.

Mass production of automobiles took place in the US, spearheaded by Henry Ford (1863–1947), an American businessperson. While working as an engineer for the Edison Lighting Company, Ford completed an internal combustion engine in 1890 and assembled a car in 1892. In 1903, Ford commenced production of the Model T, depicted in Figure 5.8, which became

FIGURE 5.7 Benz car was built in 1885.[18] (Courtesy of https://commons.wikimedia.org/wiki/File:1885Benz.jpg)

FIGURE 5.8 1910 Ford Motor Company Model T.[19] (Courtesy of https://en.wikipedia.org/wiki/Ford_Model_T)

the world's first mass-produced car. The Model T was assembled using a conveyor belt method, enabling efficient mass production, and leading to a significant reduction in manufacturing costs. The introduction of the Model T marked the beginning of the era of automobile popularization.[19] From its production launch in 1908 until 1927, over 15 million units of the Model T were sold.

5.4.2 Steel Revolutionized Railway Transportation

A transportation revolution took place during the Industrial Revolution with the adoption of iron and steel for railways, locomotives, passenger cars, and freight cars. The first wooden rail was built in Germany in the early 1550s, but it was eventually replaced by iron rails and wheels in the 1770s. These iron rails quickly spread throughout Europe. Rail transportation offered faster and safer travel compared to other means such as cars, with reduced carbon dioxide emissions due to lower friction between the rails and wheels. More than 90% of a train's weight consists of metal, predominantly steel.[20] Essential components of trains, including wheels, axles, bearings, motors, and most wagons, are made of iron and steel. Steel is also extensively used in the construction of bridges, tunnels, gas stations, train stations, ports, and airports. Additionally, steel is employed in the landing gear of aircraft, which experiences the highest momentary load.

FIGURE 5.9 Englishman Stephenson's Manchester–Liverpool steam locomotive.[22] (Courtesy of https://en.m.wikipedia.org/wiki/File:First_passenger_railway_1830.jpg)

In 1804, Richard Trevithick invented the steam locomotive.[12] In 1812, a cogwheel-driven train called the Salamanca was introduced for commercial operation.[21] In 1821, Julius Griffith, patented the first passenger locomotive, and in September 1825, the Stockton and Darlington Railway began transporting goods and passengers using a locomotive designed by George Stephenson (1781–1848). Stephenson also established a track standard of 1,435 mm width. The world's first regular train service, depicted in Figure 5.9, commenced in 1830, covering a distance of 50 km between Liverpool and Manchester, England.[22]

John Stevens (1749–1838), often referred to as the Father of American Railroads, demonstrated the potential of a steam locomotive by laying tracks on his estate in Hoboken, Northeast New Jersey, in 1826, even before regular train service began operating in England.[23] He obtained permission to lay the first railroad in North America in 1815, although it was others who pursued the construction of the first commercial railroad.

In 1830, American inventor Peter Cooper (1791–1883) designed and built the Tom Thumb, America's first steam locomotive.[24]

Edge rails, similar to modern rails, emerged in England in 1789. They consisted of single rails, approximately 1.5 m long, without flanges (protruding parts that engage with the wheels), and were attached to transverse beams and sleepers. However, they were made of cast iron, making them fragile and prone to rapid wear. The invention of the rail rolling mill in 1828 allowed for longer rails, extending to over 7 m.[25]

Railroads were established in the United States in 1827, France in 1829,[26] and Germany in 1835.[27] By 1835, the length of railroads in the United States exceeded 1,600 km, and by the 1850s, railroads had been laid in every state east of the Mississippi River. The construction of railroads in the United States is further detailed in Section 4.2.2.[28] In the late 1860s, the availability of inexpensive steel triggered a railroad construction boom not only in the United States but also in many other countries worldwide. Advancements in the steelmaking process reduced the cost of steel production, and by the end of the 19th century, most major American cities were connected by rail.

Meanwhile, Russia embarked on significant railroad construction in 1872, starting with the opening of the Trivisi to Poti line in what is now Georgia. By 1880, tracks had been extended to various parts of Central Asia, and in 1887, they reached Samarkand. The construction of the Trans-Siberian Railroad, funded by the French capital, began in 1891.[29] Despite the challenges posed by the Russo–Japanese War (1904–1905) and the Bolshevik Revolution (1917), the construction of the Siberian Railway continued, and in 1913, the world's longest line, spanning over 9,000 km, was completed. The Trans-Siberian Railway played a crucial role in stimulating the development of mines in the regions it traversed, laying the foundation for Russian industrial growth. However, the workers involved in this arduous project were not treated adequately, leading to discontent and the beginning of the Russian Red Revolution.

In China, the Wusong to Shanghai railway was initially opened by the British in 1876, but it was later dismantled due to disputes over the legitimacy of the permit.[30] Following the Sino–Japanese War, the right to build the railway was transferred to Japan and Russia, but progress was sluggish, primarily due to the targeted attacks by the Righteous Harmony Society. However, Japan recognized the importance of railways early on and actively promoted their implementation. In 1872, with the assistance of England, the section between Tokyo Shimbashi and Yokohama was opened.[31] Around 1890, the Tokaido Line, utilizing advanced technology, was completed, stretching approximately 600 km from Shimbashi to Kobe. It was also during this time that privately owned railway companies began to emerge.

In the 20th century, railways across the globe experienced remarkable growth. The traditional steam-powered trains, notorious for emitting black smoke, saw advancements in their power systems with the introduction of generators in 1866. Pollution-free electric locomotives emerged in Germany in 1881, followed by their appearance in England in 1883 and rapid adoption in major cities in the United States by 1888. The diesel engine, developed by Rudolf Diesel, a French–German inventor, in 1892, took some time to become practical for use in locomotives, with its widespread implementation occurring in the 1920s.[32] Diesel locomotives gained popularity in countries with extensive long-distance routes, such as the United States and Russia, due to their fuel efficiency and the avoidance of complex systems like power cables, poles, and transformers. Subsequent advancements were made in the efficiency of diesel-electric locomotives and rechargeable powertrains.

Beginning in the 1960s, scientists and engineers started focusing on the development of high-speed trains. From the 1970s onwards, magnetic levitation railways, which utilize electromagnetic force to levitate vehicles, garnered attention. However, due to various challenges associated with the magnetic system, high-speed railways primarily rely on electricity as their main power source. Japan's Shinkansen, which opened in 1964, was the first high-speed railway, boasting speeds of over 200 km/h upon its debut. Subsequently, high-speed trains capable of reaching speeds exceeding 300 km/h, such as France's TGV, were constructed in countries including Spain, Germany, Italy, Scandinavia, Belgium, Korea, China, England, and Taiwan. These high-speed rail systems formed the backbone of public railway transportation.[33] Notably, China's recent development of high-speed rail is noteworthy, with the country having 25,000 km of high-speed rail in 2017, accounting for 66% of the world's total. The high-speed railway in China reached 38,000 km in 2020 and is planned to expand to 70,000 km by 2035.[34] China has demonstrated unrivaled expertise in bridge construction, which is crucial for high-speed rail projects.

5.4.3 Wood to Steel: The Main Body of the Ship

Iron also revolutionized the construction of ships, which have served as a primary mode of transportation for a long time. The oldest surviving

FIGURE 5.10 The oldest Pesse Canoe (8040–7510 BC). (Courtesy of Wikimedia Commons, https://commons.wikimedia.org/wiki/File:Boomstamkano_van_Pesse,_Drents_Museum,_1955-VIII-2.jpg)

ancient vessel is the Pesse Canoe shown in Figure 5.10, dating back approximately 10,000 years, displayed at the Drents Museum in the Netherlands.(35) From the Early Bronze Age to the Middle Ages, shipwrecks discovered were predominantly made of wood. However, metal components such as copper locks have been found on ancient Egyptian ships,(36) and iron has been used for various parts like anchors, locks, and shackles since ancient Greek times.(37) With the development of iron-age civilizations, the use of metal parts in ship hulls gradually increased. Ships incorporating iron components were better equipped to withstand the forces exerted by waves and swells on the hull while occupying less space.

The advent of steam-powered ships marked another significant milestone, initially in the form of paddle steamers and later in various configurations such as composite engines with iron boilers and heating sources. One notable steamship was the Charlotte Dundas, depicted in Figure 5.11, which was constructed in 1802 by Scottish engineer William Symington (1763–1831).(38) In 1807, American inventor Robert Fulton (1765–1815) successfully built the North River Steamboat, utilizing James Watt's engine for commercial purposes.(39)

In 1819, the steamship NS Savannah successfully crossed the Atlantic Ocean for the first time, marking a significant achievement in steamship travel. This led to the emergence of numerous large passenger ships that traversed between the United States and Britain.(40) Figure 5.12 depicts the Titanic, which tragically sank on its maiden voyage in the Atlantic Ocean. The Titanic, renowned for being the largest ship in the world during its

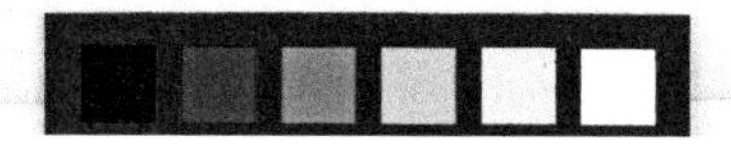

FIGURE 5.11 Steamboat Charlotte Dundas. (Courtesy of Wikimedia Commons, https://commons.wikimedia.org/wiki/File:B_Woodcroft_1846_-_The_Charlotte_Dundas_1803.jpg)

FIGURE 5.12 The Titanic was the world's largest steamship in the early 20th century. (Courtesy of https://en.m.wikipedia.org/wiki/File:Titanic-Cobh-Harbour-1912.JPG)

construction, boasted a displacement of 52,310 tons, a length of 269 m, and 11 decks. It was famously referred to as the "unsinkable" ship due to the advanced technologies of the time, including a double bottom, 16 watertight compartments, and automatically closing doors triggered by a certain water level.[41] Despite these innovations, the Titanic collided with an iceberg on April 14, 1912, during its journey in the North Atlantic Ocean and tragically sank after 2 hours and 40 minutes.

Why did the Titanic sink so quickly? Titanic was discovered in 1985 at a depth of 3,800 m, with the bow and the stern separated by about 800 m.[42] Due to the deep depth, hull sampling was very difficult. A limited number of hull pieces and rivets removed from the ship were used to make the necessary analysis. It has been reported the high sulfur content could cause brittleness at a cold seawater temperature of about –2°C, and the fracture of the hull steel was the cause of the sinking.[43] However, as a result of the investigation of the Titanic sinking into the deep sea, the range of damage to the bow collided with the iceberg was not wide, and there was no clear explanation for the phenomenon of being separated into two pieces.[44]

In 1998, Foecke analyzed two rivets recovered from the Titanic hull. The rivet material found at the fracture point was wrought iron, which contained iron silicate inclusions at a volume three times higher than normal rivets, measuring approximately 9.3% ± 0.3%. Furthermore, the microstructure near the interface where the rivet head detached (as shown in Figure 5.13) exhibited a vertical orientation with the load. This orientation made the hull prone to cracking under longitudinal load, particularly exacerbated by lower temperatures.[44] The Titanic's hull was assembled using approximately 3 million rivets.

Foecke believed that two samples were insufficient to represent the whole. Over the next 10 years, 48 more rivets were collected, and several researchers analyzed them to obtain similar results.[45, 46] Prior to Titanic, her sister ship, the RMS Olympic, entered commercial service and retired in 1935. The sister ship is that with the same design and materials. Photographs of the RMS Olympic as shown in Figure 5.14, taken after it collided with the HMS Hawke in 1911, clearly show dozens of vacant holes in the hull from which rivets popped. Foecke argues that this phenomenon may have been reproduced after the Titanic collided with the iceberg. It is considered that Foecke's argument was valid since defective rivets with high inclusion content may have accelerated the tearing of the Titanic's hull after impact. Many researchers strongly recommended the

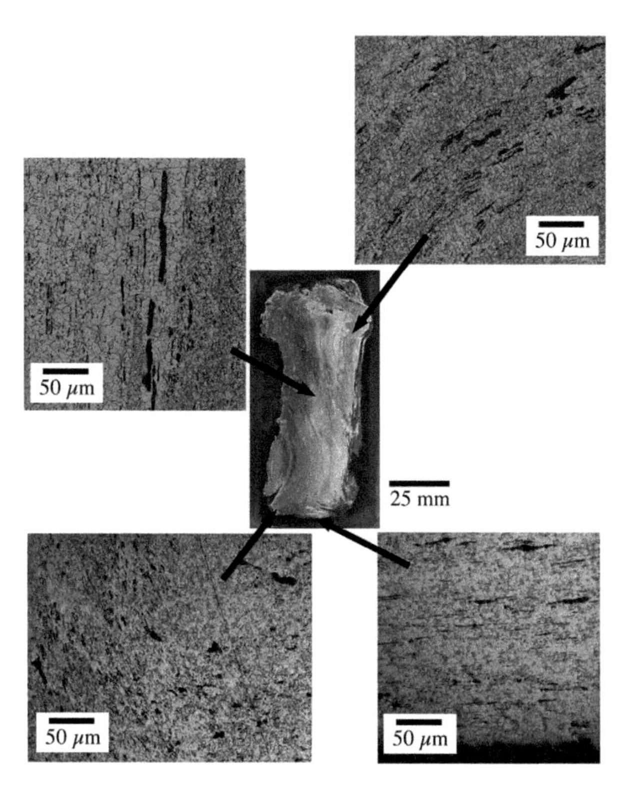

FIGURE 5.13 Montage of micrographs showing the orientation of silicate slag at various locations within a cross-section of a Titanic hull rivet.[44] (With permission from NIST https://tsapps.nist.gov/publication/get_pdf.cfm?pub_id=852863)

use of steel as a rivet material.[47] Shipbuilders of the day were moving from iron to stronger steel rivets.

The 1910s were a time when Bessemer converters were not yet available for the mass production of molten steel. As explained in Chapter 3, the steelmaking methods of the era, such as crucible and puddling, had low productivity, and wrought iron was commonly used for commercial purposes. However, if steel rivets had been used to join the Titanic's hull, which carried a large number of people, cracks would not have formed easily upon collision with an iceberg, or the propagation speed of cracks would have been slowed down. Consequently, the time until the sinking of the Titanic would have been extended. The sinking of the Titanic serves as a significant example of how much better the mechanical properties of steel are compared to those of iron.

The use of steel plates for ship hulls emerged during the American Civil War in the 1860s. Technologies for producing steel, rolling it into plates,

FIGURE 5.14 A close-up of the damage to the RMS Olympic in 1911. Note missing rivets. (Courtesy of Wikipedia, https://en.wikipedia.org/wiki/File:Olympic_Hawke_collision_damage.jpg)

cutting the plates, and drilling holes in them were developed in the 1850s. Since then, a substantial number of ships have been constructed using steel. Even the masts of ships were made of steel. Steel hulls became preferred due to their durability, ability to carry more cargo, and resistance to damage in accidents or collisions compared to wooden hulls. Moreover, a significant change occurred in the method of joining steel plates during ship manufacturing, with welding replacing riveting as the primary method of hull assembly.(48)

Steel ships accounted for 85% of global freight traffic in 2010.(49) Unlike regular ship hulls, special-purpose steel is required for ships and offshore floating bodies used for specific purposes. A notable example is the LNG carrier depicted in Figure 5.15. These carriers require stable tanks for storing liquid LNG at temperatures below −165°C. To achieve this, materials such as 304 stainless steel and 9% Ni steel have been utilized. However, a high Mn steel has recently been commercialized, which is a more cost-effective option and exhibits comparable low-temperature toughness. Consequently, an increasing number of customers are adopting this steel for their LNG carriers.(50)

FIGURE 5.15 Large LNG carrier New Apex made of steel. (With permission from Pan Ocean.)

5.5 STEEL BUILDING AND BRIDGES: HIGHER, LONGER, AND FASTER

5.5.1 The Empire State Building was the Tallest Building

The Empire State Building (ESB) shown in Figure 5.16 in Manhattan, New York, is a monumental steel-framed building that was the tallest in the world from 1931 to 1972, as many students around the world have learned. The 102-story building stands at a height of 381 m, and it reached a height of 443.2 m in 1950–1951 when a 62.2 m television transmission tower was added on top.

Since its completion in 1931, the ESB has become an iconic symbol of skyscrapers worldwide and has appeared in the background of many films such as "King Kong," "Love Affair," and "Sleepless in Seattle." Even today, with 65 elevators, the ESB continues to attract 3.5 million visitors annually. In 1945, a significant incident occurred when an American B-25 bomber crashed into the 79th floor of the ESB. Despite the seriousness of the accident, with 14 fatalities and approximately 30 injuries, the building remained standing. It is thanks to the strength of steel that the building could withstand such a collision. Approximately 60,000 tons of steel were used in the construction of this iconic structure.(51)

FIGURE 5.16 View of ESB in Manhattan, NY (left) and gilded ceiling first-floor lobby (right). (left, courtesy of Wikimedia Commons, https://commons.wikimedia.org/wiki/File:Empire_State_Building_(aerial_view).jpg, right, courtesy of https://commons.wikimedia.org/wiki/File:EmpireStateBuilding2019LobbyLookingWest.jpg)

It held the title of the world's tallest building for 41 years until the completion of the World Trade Center (WTC) in 1972. Following the tragic events of the WTC attack in 2001, the Empire State Building once again became the tallest building in New York. However, in 2015, the One World Trade Center (OWTC) was constructed in the position of the WTC, surpassing the Empire State Building in height. Subsequently, in both 2015 and 2019, two taller buildings were erected in New York, surpassing the Empire State Building in height as well. Despite these developments, the Empire State Building remains an iconic symbol of New York City.

One lesser-known record of the Empire State Building is its remarkably short construction period. The construction of the building was completed in just 1 year and 45 days, starting from the setting of the tower's first steel columns on April 7, 1930, and concluding with the finished building on March 31, 1931. This rapid construction timeline, which was completed a month before the official opening ceremonies, is an unmatched feat for a project of its scale.[52] (Figure 5.17).

FIGURE 5.17 Construction of the Empire State Building. (Courtesy of https://rarehistoricalphotos.com/empire-state-building-1931/)

The symbolism of the Empire State Building extends beyond its physical presence. It is also used as an economic indicator. The Empire State Manufacturing Index, published by the New York Federal Reserve Bank, assesses the business conditions and expectations of approximately 200 manufacturers in New York. A value below "0" indicates economic contraction, while a value above "0" signifies economic expansion. Since the Empire State Manufacturing Index was released before the Manufacturing Index of the Institute for Supply Management (ISM), which reflects the manufacturing economy across the United States, it serves as an early indicator for estimating the manufacturing economy.(53)

5.5.2 World's Longest Suspension Bridge: Çanakkale Bridge

Gelibolu, situated in the Çanakkale province of western Türkiye, faces the Dardanelles Strait. Despite being only 30 km away from Çanakkale city, reaching Gelibolu by land previously required a detour of approximately 500 km around the Sea of Marmara. Although it was possible to transport vehicles by passenger boats, it was inconvenient and expensive, with a 30-minute crossing time and limited boat schedules throughout the day. However, with the completion of the Çanakkale Bridge, the world's longest suspension bridge, depicted in Figure 5.18, on March 18, 2022, the travel time between Çanakkale and Gelibolu has been significantly reduced to 40 minutes. This bridge is expected to play a crucial role in Türkiye's economic revitalization and become a prominent tourist attraction in the future.

FIGURE 5.18 1915 Çanakkale Bridge. (With permission from DL E&C.)

This bridge was completed in 4 years, which was 1 year and 7 months shorter than the target, with DL ENC in charge of the construction and steel products supplied by POSCO. The construction of the Çanakkale Bridge holds several technical records. With a length of 3,563 m and a span of 2,023 m, it surpasses the Golden Gate Bridge's 1,280 m, making it nearly twice as long. It also exceeds the previous longest bridge, Japan's Akashi Bridge, which measured 1,991 m at the time. The main tower, constructed with steel, stands at a height of 334 m, surpassing the Eiffel Tower's 303 m.

For the main cable, an ultra-high-strength steel wire with a diameter of 5.75 mm and a tensile strength of 2.0 GPa was utilized. The main cable consists of tightly woven 18,288 steel wires, with each cable having a diameter of 881 mm and the ability to support a load of 100,000 tons, equivalent to the weight of 60,000 cars. Although the cable diameter is smaller than the Golden Gate Bridge's 920 mm, the strength has been improved to bear a heavier load with less weight. The total length of the steel wire used for the main cable is 162,000 km, which corresponds to circling the earth four times (about 40,000 km each time).[54, 55] These remarkable features demonstrate how steel, as a familiar and reliable material, promises a better future with advanced technology for people worldwide.

REFERENCES

1. M. S. Birkett, "The British Iron and steel industry", *Economica*, Vol. 5, June 1922, pp. 149–161, https://www.jstor.org/stable/2547945

2. "Torricelli's experiment", last edited February 21, 2024, https://en.wikipedia.org/wiki/Evangelista_Torricelli
3. "Otto von Guericke and the horror of vacuum", *SciHi Blog*, http://scihi.org/guericke-horror-vacuum/
4. "Denis Papin", August 22, 2016, https://www.lindahall.org/about/news/scientist-of-the-day/denis-papin
5. "Thomas Savery", https://www.britannica.com/biography/Thomas-Savery
6. "The Newcomen engine", https://www.savery.co.uk/about/newcomen-engine
7. "James Watt patent 1769 No 913", https://commons.wikimedia.org/w/index.php?title=File:James_Watt_Patent_1769_No_913.pdf&page=2
8. "Industrial Revolution: Series one", *Boulton & Watt: Parts*, Vol. 12 and 13, http://webdoc.sub.gwdg.de/zdmdm/mifoguide/matthew/INDUSTRIAL_REVOLUTION_S1_p12.pdf
9. "Sun and planet gear", last edited January 13, 2024, https://en.wikipedia.org/wiki/Sun_and_planet_gear
10. "History of the Industrial Revolution", http://www.historyworld.net/wrldhis/PlainTextHistories.asp?ParagraphID=ktw
11. "Matthew Boulton – Making the steam engine business a success", http://scihi.org/matthew-boulton-steam-engine/
12. "Richard Trevithick", last edited March 16, 2024, https://en.wikipedia.org/wiki/Richard_Trevithick
13. Hrothsige Frithowulf, "Timeline of the wheel: History and invention", Timeline of the Wheel: History and Invention - Malevus
14. Stephan Lindner, "Chariots in the Eurasian steppe: A Bayesian approach to the emergence of horse-drawn transport in the early second millennium BC", *Antiquity*, Vol. 94, No. 374, 2020, pp. 361–380, https://doi.org/10.15184/aqy.2020.37
15. "The first Iron-Rimmed wheels", https://www.sutori.com/en/item/the-first-iron-rimmed-wheels-celtic-tribes-c-1000-b-c
16. Charles H. Gibbs-Smith, "Sir George Cayley: 'Father of aerial navigation' (1773–1857), Notes and Records of the Royal Society of London", Vol. 17, No. 1, May 1962, pp. 36–56, https://www.jstor.org/stable/531013
17. "John Dunlop, Charles Goodyear, and the history of tires", https://www.cewheelsinc.com/john-dunlop-charles-goodyear-history-tires/
18. "Benz Patent-Motorwagen", last edited March 28, 2024, https://en.wikipedia.org/wiki/Benz_Patent-Motorwagen
19. "Ford Model T", last edited March 31, 2024, https://en.wikipedia.org/wiki/Ford_Model_T
20. "Establishment of an eco-friendly railway vehicle ecosystem through a virtuous cycle of materials" (in Korean), https://tech.hyundai-rotem.com/green/sustainable-railway-ecosystem-by-making-a-virtuous-circle-in-recycling-trains/
21. "Salamanca (locomotive)", last edited October 4, 2023, https://en.wikipedia.org/wiki/Salamanca_(locomotive)

22. "George Stephenson", last edited March 26, 2024, https://en.wikipedia.org/wiki/George_Stephenson
23. "John Stevens", last edited December 30, 2023, https://en.wikipedia.org/wiki/John_Stevens_(inventor,_born_1749)
24. "Tom Thumb (locomotive)", last edited February 16, 2024, https://en.wikipedia.org/wiki/Tom_Thumb_(locomotive)
25. "Rails: History and the indispensable connection with the steel industry", https://metinvestholding.com/en/media/news/reljsi-istoriya-i-nerazrivnaya-svyazj-s-metallurgiej
26. "Train", last edited March 1, 2024, https://en.wikipedia.org/wiki/Train
27. "History of rail transport in Germany", last edited March 28, 2024, https://en.wikipedia.org/wiki/History_of_rail_transport_in_Germany
28. "First transcontinental railroad", last edited March 30, 2024, https://en.wikipedia.org/wiki/First_transcontinental_railroad
29. "Trans-Siberian railway", last edited March 25, 2024, https://en.wikipedia.org/wiki/Trans-Siberian_Railway
30. "China railway history (1865–1949): First stage", https://www.travelchinaguide.com/china-trains/railway/history2.htm
31. "History of rail transport in Japan", last edited February 16, 2024, https://en.wikipedia.org/wiki/History_of_rail_transport_in_Japan
32. "Diesel locomotive", last edited March 20, 2024, https://en.wikipedia.org/wiki/Diesel_locomotive
33. Richard Nunno, "Fact sheet | high speed rail development worldwide", https://www.eesi.org/papers/view/fact-sheet-high-speed-rail-development-worldwide
34. "High-speed rail in China", last edited March 23, 2024, https://en.wikipedia.org/wiki/High-speed_rail_in_China
35. "The pesse canoe", https://drentsmuseum.nl/en/in-the-spotlight-top-exhibits/pesse-canoe
36. "Ancient Egyptians used metal in wooden ships", https://phys.org/news/2016-08-ancient-egyptians-metal-wooden-ships.html
37. Naval sciences, iconographic encyclopædia, https://www.c82.net/iconography/naval-sciences#top
38. "Charlotte Dundas", last edited November 26, 2023, https://en.wikipedia.org/wiki/Charlotte_Dundas
39. "North river steamboat", last edited August 18, 2023, https://en.wikipedia.org/wiki/North_River_Steamboat
40. "NS Savannah", last edited December 5, 2023, https://en.wikipedia.org/wiki/NS_Savannah
41. "Titanic", last edited March 26, 2024, https://en.wikipedia.org/wiki/Titanic
42. Henry Austin, "First full-size scan reveals Titanic wreck as never seen before", May 18, 2023, https://www.nbcnews.com/news/world/titanic-3d-digital-scan-wreck-sank-1912-southampton-new-york-rcna84828
43. Robert Gannon, "What Really Sank the Titanic." *Popular Science*, vol. 246, no. 2 (February 1995), pp. 49–55

44. T. Foecke, "Metallurgy of the RMS Titanic", NIST Interagency/Internal Report (NISTIR), National Institute of Standards and Technology, Gaithersburg, MD, 1998, https://tsapps.nist.gov/publication/get_pdf.cfm?pub_id=852863
45. J. J. Hooper McCarty, T.P. Weihs, and T. Foecke, "Microscopic analysis of wrought iron recovered from the wreck of RMS titanic", *Infocus*, No. 7, September 2007, pp. 82–87, doi:10.22443/rms.inf.1.26
46. J. J. Hooper, Tim Foecke, Lori Graham, and Timothy P. Weihs, "Metallurgical analysis of wrought iron from the RMS titanic", *Mar. Technol. Sname*, Vol. 40, No. 2, pp. 73–81, Paper Number: SNAME-MTSN-2003-40-2-73, https://doi.org/10.5957/mt1.2003.40.2.73
47. William J. Broad, "In weak rivets, a possible key to Titanic's doom", *The New York Times*, April 15, 2008, https://www.nytimes.com/2008/04/15/science/15titanic.html
48. Bob Irving, "Welding's vital part in major American historical events", https://www.aws.org/resources/detail/blockbuster-events-in-welding-history
49. L. M. Martinez, Jari Kauppila, and Marie Castaing, "International freight and related carbon dioxide emissions by 2050", *Trans. Res. Rec. J. Transp. Res. Board*, Vol. 2477, 2015, pp. 58–67, doi:10.3141/2477-07
50. "World's first high manganese steel LNG fuel tank installed on ultra-large crude oil carrier", https://www.marineinsight.com/shipping-news/worlds-first-high-manganese-steel-lng-fuel-tank-installed-on-ultra-large-crude-oil-carrier/
51. "Empire state building", https://www.pbs.org/wgbh/buildingbig/wonder/structure/empire_state.html
52. "Empire state building fact sheet", https://www.esbnyc.com/sites/default/files/esb_fact_sheet_4_9_14_4.pdf
53. "U.S. NY empire state manufacturing index", https://www.investing.com/economic-calendar/ny-empire-state-manufacturing-index-323
54. "1915 Çanakkale Bridge", last edited February 6, 2024, https://en.wikipedia.org/"wiki/1915_%C3%87anakkale_Bridge
55. "World's longest suspension bridge will connect Europe and Asia", https://newsroom.posco.com/en/worldsteel-worlds-longest-suspension-bridge-will-connect-europe-and-asia/

CHAPTER 6

War and Steel

SINCE THE IRON AGE, the outcome of wars has been heavily influenced by the power and effectiveness of the weapons and equipment used by soldiers. Iron played a crucial role in enhancing the strength of these weapons, making them formidable tools for both offense and defense. Consequently, nations that possessed abundant iron resources and utilized them effectively for weapon production often emerged as great powers. With advancements in the properties of iron and the development of new processing technologies, the performance of weapons continued to improve, giving rise to new and more advanced armaments. Iron, which initially found its use in small arrowheads, has undergone a continuous evolution from swords and spears to firearms, cannons, warships, submarines, aircraft carriers, and missiles.

Human history has been intertwined with warfare, which has played a significant role in driving the development of iron and steel technology. American journalist Chris Hedges defined war as a conflict resulting in the loss of more than 1,000 lives.(1) It is astonishing to note that out of the past 3,400 years, only 268 years have been marked by complete peace. Historians Richard A. Gabriel and Karen S. Metz assert that the period spanning from 1500 BC to AD 100 witnessed a genuine revolution across various aspects of human existence and organization, including the conduct of warfare.(2)

Even after the early Iron Age, conflicts persisted. Nations, both large and small, rose and fell, being annihilated by other empires or succumbing to occupation. Monarchs, aiming to prolong their power, frequently engaged in invasions of neighboring countries, and ushered in new and

DOI: 10.1201/9781003419259-6

more devastating forms of warfare in the process of defending their territories. Citizens were mobilized into the military through various means, such as ideological utopias and religious motivations. The cutting-edge technology of those eras was harnessed for weapon development, as victory or defeat in war determined the fate of nations and groups.[(3)] Based on the nature of war and the types of weapons employed, the developmental process can be divided into four distinct stages.

The first period spans from the primitive age until the invention of gunpowder. During this time, battles involved the use of offensive and defensive weapons such as swords, spears, axes, shields, longbows for long-distance attacks, and siege craft equipment such as carts and stone cannons, as depicted in Figure 6.1.[(4)] Chariots, which enabled swift movement and attacks, emerged, and technologies for constructing defensive castles were developed.

The second period covers the time from the invention of gunpowder in the ninth century until the end of the nineteenth century. Gunpowder originated in China and spread to the Middle East and Europe. The invention of firearms, credited to Berthold Schwarz in 1331,[(5)] marked a significant development. Firearms capable of long-range and high-speed

FIGURE 6.1 The most powerful siege tower in the Iron Age.[(4)] (Courtesy of https://en.wikipedia.org/wiki/Siege_tower#/media/File:Siege_of_Lisbon_by_Roque_Gameiro.jpg)

projectiles were introduced. The first recorded use of firearms in warfare occurred at the Battle of Crécy in 1346.[6] This period also witnessed advancements such as intricate muskets, butt-loading muskets, and the Gatling gun,[7] which played a role in the American Civil War in 1862.

The third period encompasses the First and Second World Wars, during which the scale of warfare expanded to include all-out conflicts between coalitions of nations, and the scope of war rapidly extended to the skies. In World War I, airplanes, airships, tanks, and submarines first made their appearance. World War II saw the practical implementation of electronic equipment such as radar, sonar, and long-range V-2 missiles. The importance of communication and security escalated, leading to the use of computer concepts for password cracking. Many of these technologies are now widely utilized in the field of information and communication technologies, including the Internet.

The fourth period spans from the end of World War II to the present day. During this time, significant advancements in warfare technology have been witnessed. Nuclear weapons, including atomic bombs and hydrogen bombs, capable of causing widespread destruction have been developed and deployed as Intercontinental Ballistic Missiles (ICBMs). Anti-Ballistic Missile (ABM) systems have also been deployed to counter potential threats to these weapons. In the field of aviation, the hypersonic era has been achieved, and Air-Launched Cruise Missiles (ALCMs) have been put into practical use. Additionally, the deployment of nuclear-powered submarines and the development of Submarine-Launched Ballistic Missile (SLBM) systems have taken place. The utilization of artificial satellites in space warfare has become prevalent. Satellites are used for various purposes such as reconnaissance, early warning systems, communication, and weather forecasting. However, there has also been a development of anti-satellite weapons known as killer satellites, which are designed to destroy enemy satellites. Other notable advancements include the development and deployment of high-power lasers, particle beams, and unmanned aerial vehicles (UAVs) in military operations. These technologies have contributed to the changing landscape of warfare in the modern era.

6.1 IRON, A KEY MATERIAL FOR ANCIENT WARFARE

6.1.1 The Hittite Empire, Pioneer of Iron Weapons

Around 1800 BC, a group known as the Chalybes, residing near the Black Sea, developed a technique for smelting iron, a metal superior in strength to bronze.[8] They used this iron to craft weapons and tools, specifically employing bloomery iron. Recognizing the value of this innovative technology, the Hittites actively embraced iron smelting methods pioneered by the Chalybes. The iron swords, spears, and chariots wielded by the Hittites outmatched any weaponry previously seen in state arsenals.[2, 9]

Equipped with iron weapons, particularly their formidable chariots capable of carrying three soldiers, the soldiers of the Hittite Empire achieved victories over their adversaries. Neighboring nations, upon witnessing the power of iron weaponry, also sought to acquire such arms. Following the decline of the Hittite Empire, knowledge of their iron manufacturing techniques spread to neighboring regions. Among the countries eager to adopt this technology, Assyria played a prominent role, utilizing it to overcome invasions by the Sea Peoples.[10]

The Battle of Kadesh: Clash between the Bronze and Iron Ages:
In 1288 BC, Egyptian Pharaoh Ramesses II and Hittite King Muwatalli II engaged in a fierce battle for control of Kadesh (depicted in Figure 6.2), a pivotal trade and strategic location. The chariot emerged as the most formidable weapon on the battlefield during this era. The Egyptian chariot carried two individuals, while the Hittite chariot accommodated three soldiers: a driver, a defender, and an attacker as shown in Figure 6.3. Reinforced with iron, the wheels of the Hittite chariot (as shown in Figure 4.3) allowed for the transportation of more soldiers, enabling efficient maneuvering and simultaneous attack and defense. The Hittites had demonstrated the superiority of their chariots in previous battles against Babylonians and other foes.[9]

Ramesses' army fell into an ambush set by the Hittites on the outskirts of Kadesh. According to contemporary Egyptian records chronicling the battle, Ramesses personally traversed the battlefield to overcome the disadvantageous situation. It is said that he crushed the besieging Hittite army. However, Hittite records indicate that Ramesses managed to escape the encirclement. Shortly thereafter, Egyptian reinforcements arrived, leading to a stalemate and ultimately the retreat of both sides. Ramesses erected a monument in his homeland proclaiming a great victory. Both Egyptians and Hittites laid claim to triumph in the Battle of Kadesh.

The Battle of Kadesh marked the first large-scale international conflict where tens of thousands of troops and numerous chariots clashed. Military experts generally acknowledge the Hittites' initial advantage over the Egyptians, as the Hittites maintained control over Kadesh following the battle.[11] After 15 years of warfare, Egypt and the Hittites entered into a

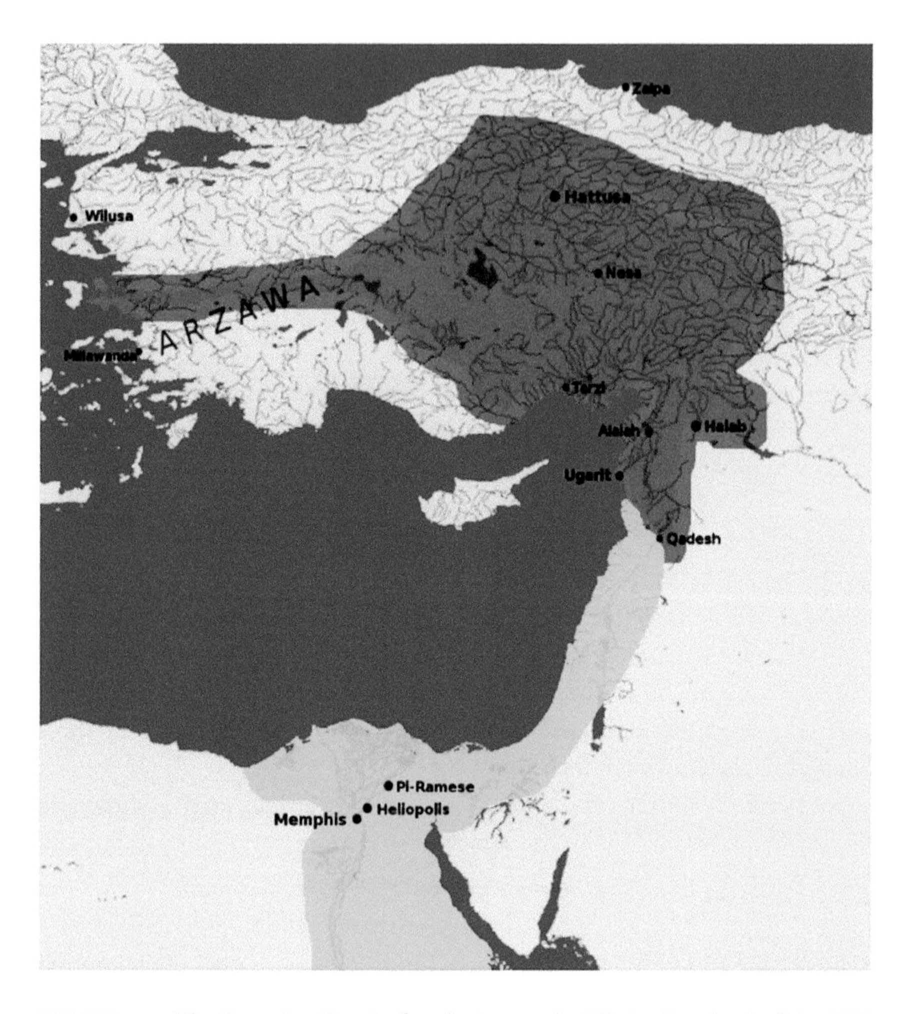

FIGURE 6.2 The Egyptian Empire bordering on the Hittite Empire (red) in 1279 BC. Kadesh is marked as Qadesh. (Courtesy of https://commons.wikimedia.org/wiki/File:Hitt_Egypt_Perseus.png)

peace treaty. In 1246 BC, Ramesses II sealed the peace by marrying Hittite King Hattusilis III's eldest daughter, Maat-Hor-Nefersureh. This period of tranquility endured until Ramesses II's death. A replica of a clay tablet engraved during that time, preserved at the United Nations as shown in Figure 6.4, serves as a reminder of the first peace treaty signed between nations.[12]

6.1.2 The Assyrian Empire: A Formidable Iron Power After the Hittites

Assyria, which emerged as an independent state in 2025 BC, derived its name from Asshur, the grandson of Noah mentioned in the Bible (Genesis

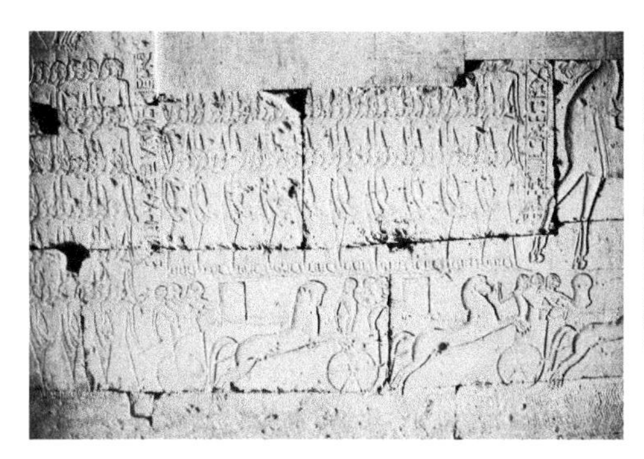

FIGURE 6.3 Comparison of armaments between Hittite (left, courtesy of © Livius.org, Kees Tol, https://vici.org/image.php?id=22758) and Egyptian chariots estimated from records and artifacts of the time (right, courtesy of https://en.wikipedia.org/wiki/Chariotry_in_ancient_Egypt#/media/File:Egyptian_War_Chariot.jpg).

FIGURE 6.4 Egyptian–Hittite peace treaty.[12] (Courtesy of https://en.wikipedia.org/wiki/Egyptian%E2%80%93Hittite_peace_treaty#/media/File:Treaty_of_Kadesh.jpg)

FIGURE 6.5 Assyrian four-seater chariot. (Courtesy of https://commons.wikimedia.org/wiki/File:P1150928_Louvre_Ninive_char_assyrien_AO19909_rwk.jpg)

10/22). The Assyrian kingdom is frequently referenced in the Old Testament and was established in the 18th century BC. However, it reached the pinnacle of its power during the Neo-Assyrian period (911 BC–605 BC). One of the key elements of Assyria's military might be its chariots, capable of accommodating four soldiers, as depicted in Figure 6.5. It should be noted that the chariot required a robust wheel to sustain high speeds while carrying such a load. The Assyrian wheel, shown in Figure 6.6, closely resembled today's wheel design, featuring eight iron spokes that ensured a balance of lightness and strength.[10]

In the Neo-Assyrian era, Assyria demonstrated its military prowess by conquering Phenicia, Israel, Babylon, and Egypt successively. Equipped with iron weapons, armor, and helmets, along with effective military strategies and bureaucratic systems, Assyrian soldiers were formidable on the battlefield. It is noteworthy that during this period, siege weapons, likely reinforced with iron, were developed, marking a significant milestone in the history of warfare. Iron helmets, resembling the Hittite helmet depicted in Figure 4.3, were also worn by Assyrian soldiers (see Figure 6.7).

Assyrian artifacts provide evidence of the widespread use of iron in their military. Soldiers and horses were equipped with iron armor, boots, daggers, spears, and arrowheads.[2] With this formidable military power,

FIGURE 6.6 Assyrian wheel, eight iron spokes mounted on a bronze rim, Chicago University.(10) (Courtesy of https://commons.wikimedia.org/wiki/File:Iron_wheels_with_bronze_hubs_-_Nabu_Temple,_Palace_of_Sargon_II,_Khorsabad,_Neo-Assyrian_period,_721-705_BC_-_Oriental_Institute_Museum,_University_of_Chicago_-_DSC07574.JPG)

Assyria expanded its dominion, establishing an empire that encompassed a vast territory in the Middle East. The empire's brutality, however, led to numerous rebellions by the conquered states, ultimately resulting in its downfall in 612 BC when a combined army composed of Neo-Babylonian, Median, Scythian, Judean, Yelamite, and Cilician forces destroyed Assyria. The last ruler of Assyria, Ashurbanipal, was renowned for establishing the world's first systematic library in Nineveh, the capital city. Many Assyrian artifacts have been discovered or are currently being excavated in Nineveh, providing valuable insights into the advanced civilization of the Assyrian Iron Age. Judging from the iron artifacts found, the peak of the Assyrian Iron Age civilization appears to have occurred in the eighth to seventh century BC. Additionally, Assyrian medical sources contain records related to iron magnetism, and some artifacts demonstrate evidence of iron welding or its combination with other materials, indicating a considerable level of iron knowledge during that time.(10)

6.1.3 Greco–Persian Wars

Following the decline of the Assyrian Empire, the Persian Empire rose to prominence in the Middle East. The Iranians, descendants of Persia,

FIGURE 6.7 Assyrian iron helmet. (Courtesy of https://en.m.wikipedia.org/wiki/File:Neo-Assyrian_Iron_Helmet,_Nimrud,_800-700_BC.jpg)

trace their origins back to Cyrus the Great, who conquered the Median and Babylonian empires and established the Achaemenid Dynasty in 550 BC. Upon the conquest of Babylonia, Cyrus issued the Cyrus Cylinder, the world's first charter of human rights, which emphasized religious freedom and abolished slavery.[13] Cyrus' son, Cambyses II, further expanded the empire by annexing Egypt, while under Darius I, Persia became an immense empire stretching from the Mediterranean to the Indus River (Figure 6.8). Darius I commissioned the construction of the Royal Road, a vital trade route connecting the eastern and western parts of the empire, and established stations along the road to facilitate the convenience and safety of merchants.[14]

As Persia expanded into Europe and Asia, it encountered resistance from the Greeks, who controlled the trade routes in the region. In response, Darius I launched an expedition in 513 BC. In the initial Persian invasion, Thrace and Macedon, some of the Greek city-states (Poleis), were occupied. Darius I aimed to conquer Athens, the leading Polis that supported the Ionian rebellion, and assembled a large fleet to cross the Aegean Sea

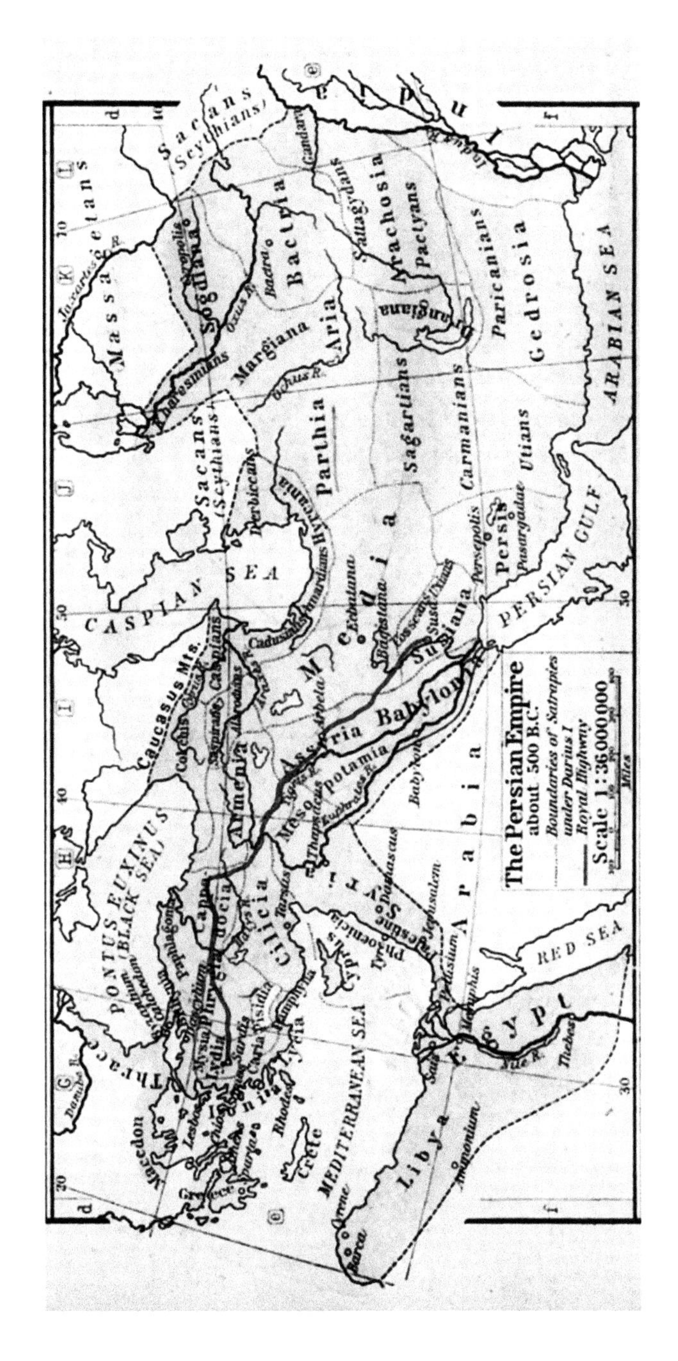

FIGURE 6.8 The Persian Achaemenid Empire at its greatest extent, 500 BC. (Courtesy of https://en.wikipedia.org/wiki/Achaemenid_Assyria)

and attack the Greek mainland. Athens, along with several other Poleis such as Sparta, allied to oppose the Persian invasion. The Greeks emerged victorious in the Battle of Marathon and Salamis, leading to the withdrawal of Persian forces. The Greek Allies, particularly Athens, which achieved victory in this war, assumed control of the East–West trade and witnessed the flourishing of a brilliant culture that influenced democratic Poleis and modern civilization. This war stands as a significant historical event that continues to shape our world today.[15]

Historians attribute the victory of the Greek allied army to several factors, including exceptional leadership, such as that of Themistocles, the coordinated teamwork of the Hoplite militia, and the individual combat skills of the soldiers. When comparing the warfare of the Persian army and the Greek Hoplites depicted in Figure 6.9, it becomes apparent that the Persians primarily relied on leather for head and body protection, while the Greek Hoplites were equipped with iron and bronze armor. This disparity in armor proved advantageous for the Greeks in close combat. Against the Persian army's arrow attacks, the Greeks could effectively defend themselves using the Phalanx strategy depicted in Figure 6.10.

FIGURE 6.9 Armament comparison of Persian soldiers on the left (Courtesy of https://commons.wikimedia.org/wiki/File:Persian_warriors_from_Berlin_Museum.jpg) and Greek Hoplites on the right (Courtesy of https://en.m.wikipedia.org/wiki/File:Ancient_athenian_warrior.jpg).

FIGURE 6.10 Greek Hoplites armaments and Phalanx tactics. (Courtesy of https://commons.wikimedia.org/wiki/File:Greek_Phalanx.jpg)

These armaments and Phalanx tactics later formed the foundation of warfare and tactics in the Roman Empire.[2]

Although Persia was not technologically inferior to Greece in terms of iron smelting, it struggled to provide heavy armor protection to soldiers who were fatigued from long-distance expeditions.[16] The lack of an efficient supply system for heavy armor was another contributing factor to their defeat. Weakened by the loss to Greece, Persia was unable to mount an effective response to the subsequent invasion by Alexander the Great, resulting in its temporary occupation.

6.2 IRON: THE MILITARY POWER OF THE ROMAN EMPIRE, HUNS, AND ISLAM

6.2.1 Iron Warfare of the Roman Empire

The Roman Empire achieved its conquests across vast territories, ranging from southern Europe to the British Isles, thanks to its heavily armed infantry and superior tactics. During this time, Roman soldiers relied on weapons and self-defense equipment made of iron, similar to the Greek Hoplites depicted in Figure 6.11. The Roman sword, known as the "Gladius" in Latin, excelled in both cutting and stabbing. This relatively short double-edged sword, measuring about 50 cm in length, proved highly effective in close combat and outmatched enemy weapons. Instead of long spears, Roman soldiers utilized javelins with a Gladius tip. These spears were thrown when the enemy closed in within approximately 20 m, and if they hit their mark, they could penetrate even the strongest armor.[2,17]

FIGURE 6.11 Depiction of Roman soldiers heavily armed with iron weapons. (Courtesy of https://commons.wikimedia.org/wiki/File:LH-R%C3%B6mer-Milites-Bedenses.jpg)

The weaponry of the Roman army continuously evolved in response to the needs of commanders and soldiers on the battlefield. When they encountered a powerful weapon used by their adversaries, the Roman armies would incorporate it into their arsenal. Regardless of the weapons employed by their enemies, the Roman army maintained military superiority in the battles for a thousand years. Iron played a pivotal role in the military might of the Roman Empire and was readily available within its borders.[2] Throughout the existence of the empire, the Romans highly valued iron. This perception of iron by the Romans influenced certain provinces, which were still using bronze, to transition into the Iron Age.[18]

6.2.2 The Iron Stirrup, the Secret Weapon That Changed World History

6.2.2.1 The Threat of the Hunnic Empire to the Roman Empire

The Roman Empire, encompassing Rome from 27 BC under Emperor Augustus (63 BC to AD 14) until the end of the Eastern Roman Empire in 1453, faced a significant challenge from the Huns. Despite the might of the

Roman Empire, the Huns demanded a substantial annual tribute in gold from them. The Xiongnu, a cavalry people active in China and northern Eurasia, possessed such power that even the Han Dynasty, which unified China, paid them tribute. As a branch of the Xiongnu, the Huns migrated westward and launched attacks on the Germanic tribes, which in turn triggered their migration to Europe and ultimately contributed to the fall of the Western Roman Empire in 476.

Emerging from the Asian continent, the Huns gained strength in western Russia and, by the end of the 4th century, swept across the European continent by crossing the Ural Mountains. With their superior martial arts and archery skills, they invaded the territories of the Slavic and Ostrogothic peoples, establishing control over the Caspian Sea and the eastern mouth of the Rhine River. Attila (410–453), depicted in Figure 6.12, was the conquering monarch who led the Huns at the height of their power.[19]

FIGURE 6.12 European painting depicting Attila and the Huns. (Courtesy of https://commons.wikimedia.org/wiki/File:Chiesa_di_San_Zaccaria_Venezia_-_Attila_conquista_Aquileia_e_Gilio_fa_mettere_in_salvo_moglie_e_figlia_(1684_circa)_affresco_di_A._Zonca.jpg)

6.2.2.2 The Stirrup: The Secret Weapon of Nomadic Horsemen

During a period of military and economic decline in the Roman Empire, the Germanic tribes on the periphery began to observe the weakening empire. In the latter half of the 4th century, the Germanic tribes expanded their settlements toward the south and east. At the same time, driven by the Huns' invasion of the Black Sea region, they started moving into the Roman Empire in larger numbers.

A significant battle took place in the Balkans between the Visigoths, a Germanic tribe, and the Roman army. In AD 378, a group of armed Visigoths, consisting of Roman mercenaries, rebelled and captured the fortress of Adrianople. The Roman Empire suffered a devastating defeat in this battle, resulting in the death of Emperor Valens (328–378) and numerous high-ranking commanders.[20]

It was the iron stirrup depicted in Figure 6.13 that played a decisive role in the battle between the Germanic tribes and the Roman soldiers. The stirrup is an equestrian device that is attached to the saddle of a horse, allowing the rider to stand securely on the horse's back. Without stirrups, the movements of Roman cavalrymen were limited. The Roman army proved no match for the Goths, who fastened their feet to the stirrups and wielded spears and bows with greater ease and maneuverability.

FIGURE 6.13 Horsemanship using stirrups. (Courtesy of https://commons.wikimedia.org/wiki/File:Hun_warriors.JPG)

FIGURE 6.14 Goguryeo horseback archery Parthian Shot. (Courtesy of http://www.heritage.go.kr/heri/cul/culSelectDetail.do?VdkVgwKey=11,00630000,32&pageNo=5_1_1_0#)

The Goths learned how to make and utilize stirrups from the nomadic peoples.[21] The nomadic fighting technique of shooting arrows while turning around on a galloping horse was unimaginable in Europe at that time. Sidonius Apollinaris, a 5th-century politician, and bishop, described the horsemanship of the Huns, stating, "You would think the limbs of man and beast were born together, so firmly does the rider always stick to the horse."[22] The Goguryeo Muyongchong mural depicted in Figure 6.14 shows a rider twisting his body backward, pulling the bowstring to his ear, and aiming at the target.[23] This technique is known as the Parthian Shot, which originated from the mounted archery practices of northern horse-riding peoples.[24]

The Battle of Carrhae[25] in 53 BC between the Parthians and the Romans is known for the Parthian Shot. The retreating Parthian cavalry suddenly turned their upper bodies and shot arrows at the pursuing Romans. This unexpected attack confused the Roman army, and approximately 1,000 heavily armored Parthian cavalrymen rushed in, disrupting the Roman forces. With the use of iron stirrups, the Parthian cavalry could perform their skilled tactics with agility and precision, even with minimal training. Once their feet were secured in the stirrups, they could maintain a stable posture and shoot their bows in any direction without compromising their balance.

Toward the end of the 4th century, the Huns emerged as a threat to the Romans. Originating from the northern steppes of the Aral Sea, the Huns, under the leadership of Balamir, began their westward expansion around AD 374. Armed with iron weapons such as saddles, stirrups, composite bows, triangular iron arrowheads, and arrows, the Huns crossed the Volga, Don, and Dnieper rivers, overpowered the Ostrogoths, and defeated the Visigoths. Their remarkable mobility and exceptional cavalry tactics quickly overwhelmed the Germanic tribes. Figure 6.15 depicts a map showing the territory controlled by the Huns during that time.(19) To escape the Huns, the Goths crossed the Danube and sought refuge in Roman territory. Later, when Uldin, the son of Valamir, attacked the Eastern European plains, the Goths moved further into the Italian peninsula, triggering the significant historical event known as the "Migration Age."(26)

FIGURE 6.15 The Empire of the Huns and subject tribes at the time of Attila.(19) (Courtesy of https://commons.wikimedia.org/wiki/File:Huns450.png)

In 434, Attila launched indiscriminate attacks on the Germanic tribes in the surrounding areas. By 441, Attila had cleared the vicinity of Rome and declared war on the Eastern Roman Empire, crossing the Danube and causing major cities to be razed. Consequently, the Eastern Roman Emperor Theodosius (401–450) was compelled to sign a humiliating peace treaty with Attila's forces in 442.

Attila's forces once again assaulted the Eastern Roman Empire in 447, devastating the Balkans except for Constantinople. In 451, Attila turned his attention to Gaul, capturing Metz and laying siege to Orléans. The Western Roman Emperor fled, and Roman Archbishop Leo appealed for reconciliation. However, in the following year, Attila passed away on the first night of his marriage to Ildico, the daughter of a German prince. After Attila's death, the Hun Empire weakened as his three sons contended for the throne.(19)

6.2.3 Damascus Sword and Other Legendary Swords

6.2.3.1 Damascus Sword, a Deadly Islamic Weapon

Arabs who embraced Islam emerged from the Arabian Desert, conquering North Africa and governing Spain on the Iberian Peninsula for 700 years. Islamic science and technology, which were more advanced than those in Europe, played a significant role in their overwhelming power. In the late 11th century, the Seljuk Turks captured Palestine and Syria, territories that had belonged to the Eastern Roman Empire. As the threat to Christian pilgrims visiting the region escalated, Pope Urban II called for a military campaign to address the issue in 1095, marking the beginning of the Crusades War. In the First Crusade (1096–1099), the Crusaders seized Jerusalem, the holy city, and established Christian states such as the Kingdom of Jerusalem.(27) However, it didn't take long for the formidable Crusaders to face defeat at the hands of Muslim armies. The Islamic army possessed the renowned "Damascus Sword" depicted in Figure 6.16, which was often referred to as the "Mysterious Sword."(28)

The Damascus sword is a representative sword of Islam and was mainly produced in Damascus, Syria, by importing Wootz steel from India.(29) The blade of the Crusaders' sword was thick and heavy, but the Damascus sword was thin, light, and sharp, and did not break even when it struck against rocks. The Damascus sword of the Islamic army pierced the armor of the Crusaders, who were not so mobile due to heavy weapons and equipment.

FIGURE 6.16 A genuine Damascus steel blade. (Courtesy of https://www.metmuseum.org/art/collection/search/30970)

On the surface of Damascus swords, a wave-shaped pattern called "Damascus" is spread, giving a sense of mystery. Legends are told that Richard the Lionheart of the Crusaders met King Saladin, a great Muslim hero, and was greatly surprised to see the Damascus sword cut through windblown silk.[30] As India's iron ore ran out after the 15th century, production of the Damascus sword ceased in 1750, and its manufacturing method was forgotten.

6.2.3.2 Damascus Sword, Made With Carbon Nanotube Technology

Marianne Reibold and colleagues from the University of Dresden uncovered the extraordinary secret of Damascus steel–carbon nanotubes (CNT). The smiths of old were inadvertently using nanotechnology.[31, 32] Damascus blades were forged from small cakes of steel from India called 'Wootz'. All steel is made by allowing iron with carbon to harden, resulting in a metal. The problem with steel manufacture is that high carbon contents of 1–2% certainly make the material hard but also render it brittle. This is useless for sword steel since the blade would shatter upon impact with a shield or another sword. Wootz, with its especially high carbon content of about 1.5%, should have been useless for sword-making. Nonetheless, the resulting sabers showed a seemingly impossible combination of hardness and malleability.

Reibold's team solved this paradox by analyzing a Damascus saber created by the famous blacksmith Assad Ullah in the seventeenth century, graciously donated by the Berne Historical Museum in Switzerland. Amazingly, they found that the steel contained CNTs, each one just

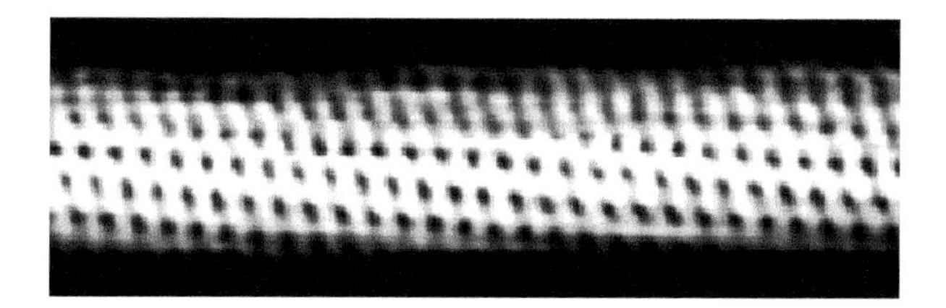

FIGURE 6.17 Structure of carbon nanotube. (Courtesy of https://en.m.wikipedia.org/wiki/File:Chiraltube.png)

slightly larger than half a nanometer. Ten million could fit side by side on the head of a thumbtack.

CNTs are cylinders made of hexagonally arranged carbon atoms, as shown in Figure 6.17. They are among the strongest materials known and possess great elasticity and tensile strength. In Reibold's analysis, CNTs were found to be protecting nanowires of cementite (Fe_3C), a hard and brittle compound formed by the iron and carbon of the steel. This composite material at the nanometer level explains the steel's special properties. The malleability of CNT compensates for the brittle nature of the cementite formed by the high-carbon Wootz cakes.

The exact process by which the ancient blacksmiths produced these nanotubes remains unclear, but researchers believe that small traces of metals in the Wootz, including vanadium, chromium, manganese, cobalt, and nickel, played a crucial role. The alternating hot and cold phases during manufacturing caused these impurities to segregate into planes. These planes then acted as catalysts for the formation of CNTs, which, in turn, promoted the formation of cementite nanowires. These structures aligned along the planes defined by the impurities, accounting for the characteristic wavy bands or damask patterns found in Damascus blades. CNT is a novel material with carbon particles arranged in a hexagonal honeycomb shape. It possesses the same thermal conductivity as diamond and is several times stronger than steel.

Comparing the mechanical properties of Tiberius,[33] the famous Roman sword Gladius mentioned in Section 4.1.3, with Damascus steel is intriguing. Tiberius is composed of martensite with a carbon content ranging from 0.4% to 0.6%, while Damascus steel contains approximately 1.5% carbon with a CNT structure that exhibits exceptional toughness. The influence of the CNT structure on the mechanical properties is not yet fully understood. However, analyzing the effect of CNT carbon on mechanical performance and applying the findings to the manufacturing of high-performance steels would be a highly challenging topic.

FIGURE 6.18 The legendary sword, Excalibur.[34] (Courtesy of https://en.wikipedia.org/wiki/Excalibur#/media/File:An_island_story;_a_child's_history_of_England_(1906)_(14801002423).jpg)

6.2.3.3 The Story of the Legendary Swords

The story of the legendary sword begins in the West with the iconic Excalibur, a sword famously depicted as embedded in a rock, as shown in Figure 6.18. King Arthur, a Celtic tribal chief believed to have lived in England during the 5th and 6th centuries, is closely associated with this legendary sword.[34] However, stories of such legendary swords have been passed down from other countries as well. This is because the sword holds symbolic significance when discussing historical figures. It served as a weapon for attacking enemies during war and as a means of self-defense in emergencies. Furthermore, it played a vital role as a tool for survival and everyday life. A legendary sword also represented the owner's social status, power, and personality.

Creating a masterful sword required the fusion of a blacksmith's refined skills and unwavering spirit. Especially in ancient and medieval times, the foundation for producing high-quality steel products was not well-established. Therefore, crafting a legendary sword of exceptional quality involved undergoing countless iterations of trial and error, as well as employing various post-processing techniques shrouded in secrecy.

In Europe, there are various legendary swords, including "Excalibur," associated with King Arthur, "Gram,"[35] wielded by Sigurd in Germanic mythology, "Harpe"[36] used by Perseus in Greek mythology, and "Ascalon"[37] from Israel. One notable famous sword that still exists is the "Gladius"[17, 33] of the Roman Empire. The length of the gladius was intentionally short to align with the fighting style of the Roman army. The Romans employed tight formations and followed well-planned strategies and rules in battles. In this context, shorter weapons demonstrated excellent effectiveness, particularly in large-scale close combat. The invaders who confronted the Roman Empire possessed longer swords, which may have surpassed Roman soldiers in individual combat abilities but proved ineffective in large-scale confrontations.

The Colosseum, a gigantic circular amphitheater built in AD 105, is a prime example of an arena where Gladiators fought, and their battles were a crucial spectacle for the Roman people that could not be missed. Gladiator fights remained popular for hundreds of years, giving rise to social issues such as the human rights of Gladiators, various interests at stake, and their exploitation as a political tool. Eventually, Gladiator fights were prohibited by imperial decree. With the decline of the Roman Empire, sword competitions and similar events vanished. The Romans referred to their performers as Gladiators, deriving the term from the root word "Gladius." The Gladius is not solely a Gladiator's weapon but a sword carrying historical symbolism that intertwined with the rise and fall of Rome.[38]

Legendary swords can be found not only in Europe but also in other parts of the world, including China, Japan, Korea, and the Middle East. In China, tales of various famous swords have emerged since the Spring and Autumn period. Examples include the swords of "Gan Jiang and Mo Ye,"[39] a legendary blacksmith couple from the Wu Dynasty; "Paekpido,"[40] used by Cao Cao in the Three Kingdoms period; and Liu Bei's "Ssanggogeom."[41] In Korea, renowned swords such as Kim Yu-sin's "Saingeom,"[42] a general of Shilla, Baekje's "Chiljido,"[43] "Jeondo,"[44] from the early Joseon Dynasty, and Yi Sun-sin's "Ssangyonggeom,"[45] wielded by the great general during the Japanese invasion of Korea in 1592, are celebrated as legendary swords. Japan, too, boasts the "Tenkagoken," a collection of five famous swords that have been preserved to this day.[46] Three of the Tenkagoken are designated National Treasures of Japan, one belongs to the Japanese royal family and another is held by a Buddhist sect as a sacred object.

6.3 HISTORY CHANGER: BATTLESHIPS AND CANNONS MADE OF IRON

6.3.1 The Basilic Cannon and the Fall of the Eastern Roman Empire

In the 4th century, Constantine the Great constructed the formidable Theodosian Wall, extending nearly 20 km along the coastline of Constantinople. Capitalizing on the strategic advantage of the narrow entrance to the Golden Horn, a massive chain similar to the one depicted in Figure 6.19 was suspended across the 750-m opening of the harbor to prevent enemy ships from infiltrating. This chain, known as the Golden Horn Chain, is documented in historical records[47] as having been used in 821, 969, 1203, and 1453. Notably, during Mehmed II's invasion of Constantinople in 1453, the Ottoman Empire could not breach the Golden

FIGURE 6.19 Installation drawing and actual chain across Golden Horn.[47] (Courtesy of https://en.wikipedia.org/wiki/File:20131205_Istanbul_136.jpg)

Horn, which was the weakest point in the Theodosian Wall, due to this chain. After much contemplation, Mehmed II made a historic and audacious attempt to transport his warships overland, across Galata Hill, and into the waters of the Golden Horn.(48) It is intriguing to explore how the chain remained intact despite being made of wrought iron and having been submerged in corrosive seawater for numerous years. It is presumed that the iron ore used in the construction of this chain possessed a chemical composition that conferred resistance to corrosion.(49)

In the Siege of Constantinople in 1453, the Basilic cannon, depicted in Figure 6.20, made its appearance.(50) This colossal cannon, crafted by Orban, a renowned cannon engineer, stood as one of the largest cannons ever constructed in history. It measured an impressive length of 8 m and possessed the capability to launch a 270 kg stone projectile up to a distance of 1.6 km, effectively demolishing the seemingly invincible walls of Theodosius. However, unleashing such a massive projectile required an immense amount of gunpowder. Once fired, the gun barrel would overheat, necessitating a three-hour cooldown period before it could be fired again.(51) This susceptibility to heat, resulting in the occurrence of cracks within the bronze gun barrel, ultimately led to the cannon's abandonment after the war. In 1807, during the Battle of Dardanelles between the Royal Navy and the Ottoman Empire, this 15th-century cannon resurfaced,

FIGURE 6.20 The Basilic cannon was built in 1464 and modeled after the Orban Cannon used in the Ottoman siege of Constantinople in 1453. British Royal Armory Collection. (Courtesy of https://en.wikipedia.org/wiki/Orban#/media/File:Dardanelles_Gun_Turkish_Bronze_15c.png)

firing upon the British fleet, and inflicting 28 casualties. Made entirely of iron, this cannon boasted a diameter of 63 cm and weighed 1,028 kg.[50]

6.3.2 Beginning of the Age of Discovery, Spain and Portugal

Britain, which developed its strength through the Industrial Revolution based on innovative iron and steelmaking technology, reigned as the world's most powerful country until the early 20th century. With colonies spread across various parts of the world, Britain earned the epithet "the country where the sun never sets" as it was daytime in its colonies in India, Southeast Asia, and North Africa even when it was nighttime in the home country. At one point, Britain ruled over a quarter of the world's land area and approximately 20% of the world's total population, as depicted in Figure 6.21.[52] During the 19th century, the British Empire leveraged its formidable naval power to secure trade rights and achieved dominance through colonization and trade.

Before the rise of the British Empire, Spain, and Portugal enjoyed a long period of power. From the late 15th century to the middle of the 18th century, known as the Age of Discovery, European ships embarked on voyages around the world, pioneering new trade routes. Christopher Columbus discovered the Americas, while Vasco da Gama established a sea route to India. Pedro Alvares Cabral discovered Brazil, and Ferdinand Magellan accomplished the historic feat of crossing the Pacific and Atlantic Oceans, as shown in Figure 6.22.[53]

European nobility had a strong affinity for spices such as pepper and nutmeg, which were primarily produced in India and transported to Europe via Türkiye. The Ottoman Empire, which controlled this trade route in the 15th century, held a monopoly over the spice trade and sold it to Europe at exorbitant prices. In response, Portugal sought an alternative route, believing that they could reach India by circumnavigating the African continent to the south and east. Conversely, Spain became interested in the Western route and dispatched Columbus to the Americas in 1492. Until the mid-16th century, when Spain and Portugal dominated international trade and controlled the seas in Asia and the Americas, England played a relatively minor role.

6.3.3 British Empire's Power: Iron and Steelmaking Technology

The dominance of Spain and Portugal was relatively short-lived. In 1588, the British defeated the Spanish Armada, which was considered the world's most powerful fleet at that time, in the Battle of Calais, as depicted

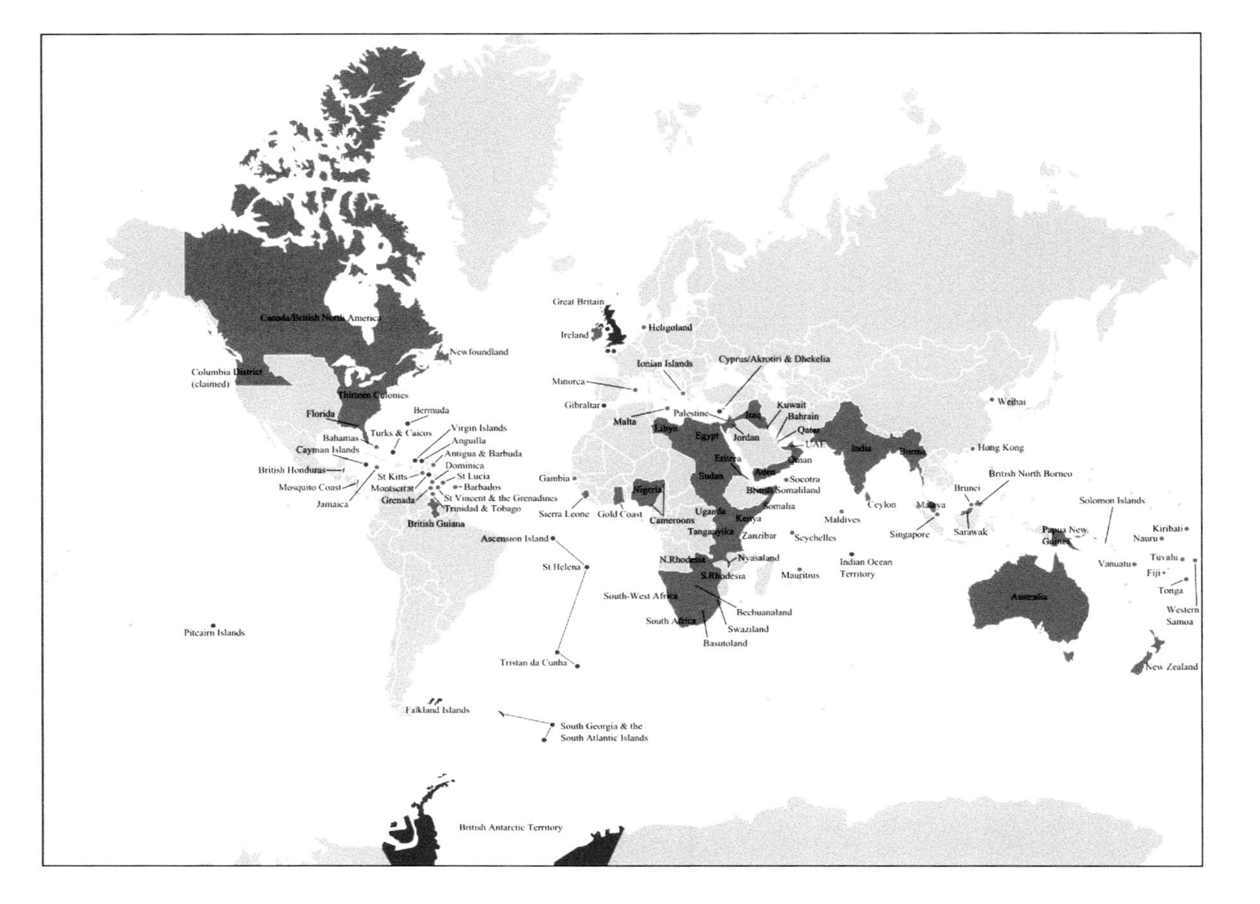

FIGURE 6.21 Territories of the British Empire in 1921.[52] (Courtesy of https://commons.wikimedia.org/wiki/File:British_Empire_Mercator.svg)

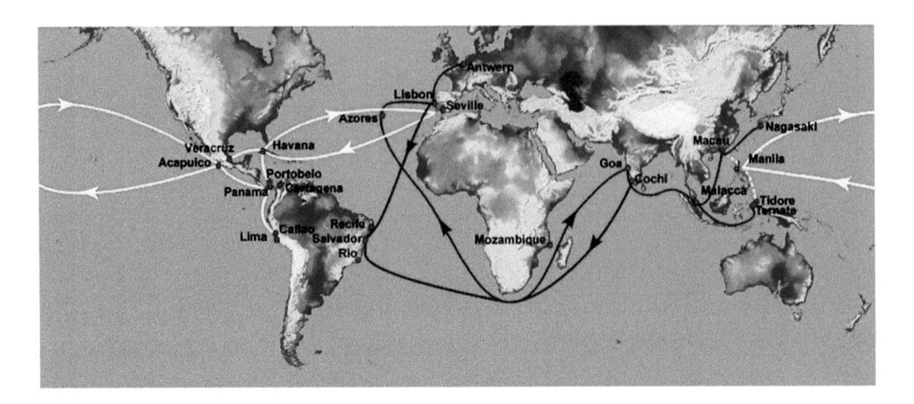

FIGURE 6.22 Trade routes pioneered by Spain and Portugal during the Age of Discovery.[53] (Courtesy of https://en.wikipedia.org/wiki/Age_of_Discovery#/media/File:16th_century_Portuguese_Spanish_trade_routes.png)

FIGURE 6.23 English fireships launched at the Spanish Armada off Calais. (Courtesy of https://commons.wikimedia.org/wiki/File:Spanish_Armada_fireships.jpg)

in Figure 6.23. The outcome of naval battles heavily relied on the maneuverability and durability of battleships, as well as the disparity in firepower between the cannons mounted on the ships and the guns carried by soldiers. During that period, Britain excelled as a technological leader in iron manufacturing and the production of iron weapons. An analysis of the opposing battleships following the Battle of Calais revealed significant technological disparities between the two countries in terms of battleships and weapon manufacturing.

It was found that no English ships had sustained hull damage, while many of the Spanish ships were severely damaged by cannon fire, particularly below the waterline. Examination of Spanish cannonballs retrieved from shipwrecks revealed that the Armada's ammunition was poorly cast, with the iron lacking the correct composition and being too brittle. As a result, the cannonballs disintegrated upon impact rather than penetrating the hull. Several guns were also found to be poorly cast and composed, increasing the risk of bursting and endangering the gun crews.[53–56]

Despite being an island nation, Britain did not have a navy. When Edward III (1312–1377) assembled a fleet of 700 ships in total to invade France in 1347, only 30 of them belonged to the king. However, under Henry VIII (1491–1547), who sought to build a strong empire, investments were made in shipyards and warships, such as the Mary Rose depicted in Figure 6.24.[57] The accomplishments of Admiral Francis Drake (1563–1596), who became the first Englishman to circumnavigate the world, combined with the ambitions of the new English trading companies, led

FIGURE 6.24 Mary Rose with many heavy guns. (Courtesy of https://commons.wikimedia.org/wiki/File:AnthonyRoll-2_Mary_Rose.jpg)

Britain to recognize the importance of maritime dominance for its future. During the Battle of Calais in 1588 against the Spanish Armada Fleet, Drake, serving as the second-in-command of the English fleet, employed a pirate-style strategy of sending a ship loaded with gunpowder and oil to set the enemy fleet on fire, creating chaos among the Armada Fleet.(54, 55)

Walter Raleigh (1554–1618), an English politician and a known favorite of Queen Elizabeth I, famously stated, "For whosoever commands the sea commands the trade; whosoever commands the trade of the world commands the riches of the world, and consequently the world itself."(58) Following Raleigh's philosophy, Britain expanded its naval forces, built more warships, and produced more iron cannons than any other country. Throughout history, the shipbuilding industry has been associated with the rise of hegemonic powers. The British dominated the world shipbuilding industry until the 1950s, particularly with the advent of steamships and iron ships. The UK developed a rivet joining technique, wherein the ends of two iron plates were overlapped, a hole was made, and a thick iron nail heated over a fire was driven in for joining. After fulfilling the demand for warships during World War I, Britain turned its attention to the global merchant and passenger ship markets.(59) This rivet joining method was eventually replaced by the welding method developed during World War II.

6.3.4 Japanese Force in the Pacific War, Armed With Huge Warships and Guns

On May 27, 1905, the Russo–Japanese War broke out between the Japanese Combined Fleet and the Russian Baltic Fleet off the coast of Tsushima Island, Japan. The war concluded swiftly with Japan achieving a unilateral victory as the bridge of the flagship of the Russian Fleet was destroyed by the guns of the Japanese warship.(60) The Japanese warships deployed for the Tsushima Naval Battle were constructed by Britain at Japan's request.(61) During British Prime Minister Theresa May's visit to Japan on August 31, 2017, she had the opportunity to board the Japan Maritime Self-Defense Force's newest helicopter carrier and flagship, the Izumo. Japanese Defense Minister Onodera explained, "The ship Izumo shares the same name as the flagship of the Imperial Japanese Navy (as shown in Figure 6.25) that destroyed the Russian fleet during the Russo–Japanese War in the past. Japan was able to achieve victory thanks to the Izumo."(62)

At the beginning of the Pacific War, Japan's preemptive attack on the United States proved highly effective. Japan managed to occupy numerous

FIGURE 6.25 Warship Izumo led the Russo–Japanese War to victory.[61] (Courtesy of https://en.m.wikipedia.org/wiki/File:Japanese_cruiser_Izumo_in_Kobe.jpg)

areas in Southeast Asia, as depicted in Figure 6.26. However, Japan's victory was short-lived. As the war progressed, the United States significantly increased steel production and mass-produced wartime supplies. Furthermore, the US possessed advanced technologies that Japan lacked, including code-breaking and radar. With its formidable air and naval

FIGURE 6.26 Areas occupied by Japan during the Pacific War. (Courtesy of https://en.m.wikipedia.org/wiki/File:Japanese_Empire_-_1942.svg)

power and unlimited resources, the US recaptured the Southeast Asian region that had been seized by Japan. The war situation shifted advantageously in favor of the US, and Japan attempted to compensate for this through the use of Kamikaze commandos. Japan's desperate resistance ultimately led to the US deploying nuclear weapons, culminating in Japan's unconditional surrender and the conclusion of the Pacific War.(63)

6.4 WORLD WARS AND STEEL

6.4.1 The Rapid Growth of the US Steel Industry during World War II

The United States significantly ramped up steel production during World War II. Tanks, cannons, warships, fighter planes, and virtually every item required for war relied on steel production. In March 1943, when the war was in full swing, Irving Sands Olds (1887–1963), the CEO of US Steel, reported that the company had produced over 30 million tons of steel in the first year of the war, setting a new production record.(64)

World War II served as a stark example of how crucial steel was for manufacturing weapons, and it marked a turning point for the American steel industry. In 1945, the US accounted for approximately 70% of the world's steel production. According to records from the US National War Museum, Bethlehem Steel (BSC) produced an astounding 73.4 million tons of steel by war's end in 1945. Apart from the steelworks in Pennsylvania, BSC had several other steel plants, employing a total of 283,765 workers.(65)

BSC secured a US$1.3 billion order for munitions just weeks after the attack on Pearl Harbor. During World War II, BSC played a crucial role in production, manufacturing 70% of aircraft cylinder forgings, 25% of battleship armor plates, and one-third of artillery forgings. BSC's contribution extended to shipbuilding as well, constructing 1,127 ships and producing various components of warships, including guns and shells as depicted in Figure 6.27. Notably, BSC accounted for 20% of the total fleet of the US Navy.(65)

According to American historian Arthur L. Herman (1956–) in his 2012 book, before World War II, the US Army had only 270,000 men in 1940, and its weapon system was lacking.(66) In the summer of 1939, just before Hitler's army invaded Poland, General George Patton (1885–1945), who would later become a renowned military commander, assumed command of an armored force comprising 325 tanks. During that time, the supply system for his unit was so deficient that he had to personally purchase the necessary bolts and nuts for his troops. However, the logistical

FIGURE 6.27 Bethlehem Steel cannon production during World War II.[65] (Courtesy of https://en.wikipedia.org/?title=File:Shop_Number_2_annex._6_inch_guns_with_their_mounts_in_foreground,_immediately_in_the_rear,_slides_or_cradles_for_10_inch_-_NARA_-_533708.tif&page=1)

system underwent significant strengthening, and the number of US troops increased to 1.46 million during World War II. In total, approximately 16 million US soldiers participated in the entire conflict.

After the unexpected Japanese attack on Pearl Harbor, President Roosevelt swiftly declared war and initiated the transition of the US economy into wartime mobilization. The nation mobilized all of its resources, including 54 million workers. Civilian volunteers were granted military-style ranks and uniforms, as depicted in Figure 6.28, and even wealthy individuals worked for their country, receiving a salary of only US$1 a year. The United States accounted for two-thirds of the weapons and equipment produced by all the Allied forces. Following the conclusion of World War II in 1945, Soviet dictator Stalin (1878–1953) openly acknowledged, "If the US had not assisted us, we would not have emerged victorious. If we had been left to face Nazi Germany alone, we would not have withstood Germany's pressure, and we would have lost the war."[67]

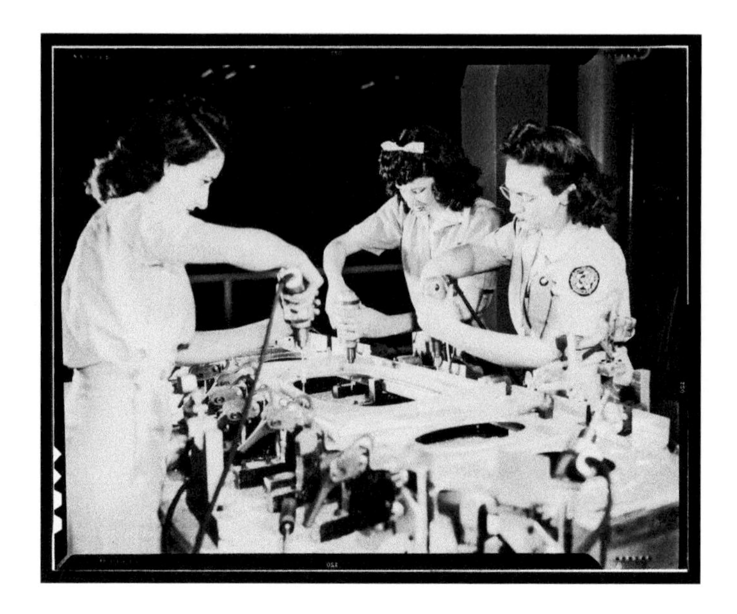

FIGURE 6.28 Women workers made airplane wings in a Texas factory during World War II. (Courtesy of https://www.loc.gov/resource/fsac.1a34931/)

Isoroku Yamamoto (1883–1943), the commander-in-chief of the Japanese Combined Fleet, recognized that the gap in steel production capacity between the US and Japan was elevenfold. He understood that engaging in a prolonged war with the US would be unfeasible due to the vast disparity in steel infrastructure between the two countries. Consequently, Yamamoto intended to weaken US naval power through a surprise air attack on Pearl Harbor, aiming to subdue the US in a diplomatic battle afterward.[(68)] Yu Tianren (俞天任), a Chinese writer residing in Japan, stated in his report titled "Advisors in Daihonai, 大本營," that during the Pacific War, the Japanese army consumed 78 million shells, while the US army used 4 billion shells.[(69)] In a conflict centered on iron and steel, Japan was outmatched by the US from the very beginning.

6.4.2 The US, a Leader of the Free Camp After the World War II

The US, as a leader of the free world, emerged prominently after World War II. One significant event that established the US as a major power was the 1898 conflict with Spain. The US emerged victorious in this war, resulting in the occupation of Spanish colonies ranging from Cuba to the Philippines. While the US economy was smaller than that of Great Britain in 1820, it caught up and became comparable to Great Britain's size by

1870. By the end of World War I in 1918, the US had become the world's largest economy.

Before embarking on full-scale industrialization, the US experienced the Civil War, which erupted due to the issue of slavery abolition. One of the factors that contributed to the victory of the Union Army in the Civil War was the development of commerce and industry in North America. These developments led to the supply of weapons such as the Parrot rifle made of cast iron having an advantage over the Confederate Army and its performance. More than a thousand new artillery pieces were produced in Union factories between 1861 and 1865, the most famous of which was the Phoenix Iron Works in Phoenixville, Pennsylvania. In particular, at Pickett's Charge in Gettysburg, which was a turning point in the Civil War, the artillery of the Union army outperformed the artillery of the Confederate army and was able to break the brunt of the Confederate army.(70) The Union's supply system for wartime materials followed a similar pattern in the subsequent world wars, fully demonstrating the potential of the US.

During World War II, the US solidified its position as the leader of the free world. Although the US actively participated in the war, its mainland remained largely untouched, and its war casualties were relatively small, totaling 300,000, which was one-eightieth of the casualties suffered by the Soviet Union. Simultaneously, the US harnessed its industrial potential and played a crucial role as the "Allied Arsenal." After the war, the US held a dominant position in various aspects. American commercial shipping and gold reserves accounted for two-thirds of the global stock. The US dollar replaced the British pound as the international reserve currency. American coal production accounted for nearly half of the world's output, while crude oil production surpassed two-thirds. Consequently, the US became the leader of the free world, capable of countering the communist Soviet bloc that emerged after the war.(71)

6.4.3 The Pride of Japan: The Battleships Yamato and Musashi

Even after the outbreak of World War II, Japanese naval military doctrine continued to emphasize the importance of "out-gunning and out-fighting the enemy" to achieve victory. However, with the onset of the Pacific War, which involved battles over vast distances spanning thousands to tens of thousands of kilometers, the significance of large ship guns rapidly diminished, and naval battles centered around aircraft carriers became the focal point of tactics. In an attempt to neutralize the military power of the US Navy, which relied heavily on aircraft carriers, Japan launched a preemptive strike on Pearl Harbor.

Despite being products of the Japanese giant warship doctrine, the battleships Yamato and Musashi did not prove effective in assisting Japanese forces in the Pacific War. These battleships were deployed to the rear and primarily served as the flagship vessels for the Commander-in-Chief of the Navy. Toward the end of the war, when Japan's aircraft carrier fleet was nearly exhausted, Yamato and Musashi were deployed as the last hope of the Japanese military. However, they were ultimately sunk in vain by the relentless attacks of American aircraft carriers. In 1953, Kitaro Matsumoto, who had been involved in engineering projects related to the design and construction of the Yamato battleship, contributed some insights to the US Naval Institute. He provided details about the steel-related aspects of the project.[72] Table 6.1 provides an overview of Yamato's final specifications, and Figure 6.29 depicts the ship under construction.

In the construction of Yamato and Musashi, butt-joining was exclusively used for the bow and stern hulls, while lap joints with rivets were employed for the rest of the hull. The total number of rivets utilized throughout the hulls amounted to 6,153,030. However, during the Battle of Seals, when Yamato was struck by a torpedo, it was evident that the side armor was relatively weak, only able to withstand penetration of approximately 1 m. The manufacture of thick steel plates, particularly those with a thickness of 400 mm or more, to protect the battleship's sides also faced limitations. At that time, Japan's steelworks and forging factories had a maximum weight capacity of 68.5 tons, and the area covered by a 410 mm

TABLE 6.1 Principal Dimensions of the Final Yamato Project Plan[72]

Length between perpendiculars, m		244
Length overall, m		263
Breadth maximum, m		38.9
Mean draft (Full load condition), m		10.86
Displacement, tons	Full load condition	69,100
	Trial conditions	72,809
Fuel load capacity, tons		6,300
Radius of action, miles at 16 knots		7,200
Shaft Horsepower, HP		150,000
Engines		4 turbines
Steam pressure, kg/cm^2 and temperature, °C		25 and 325
Number of crews		approx. 2,500

FIGURE 6.29 Yamato was under construction at Japan's Kure Naval Base on September 20, 1941. (Courtesy of https://ko.wikipedia.org/wiki/%ED%8C%8C%EC%9D%BC:Japanese_battleship_Yamato_fitting_out_at_the_Kure_Naval_Base,_Japan,_20_September_1941_(NH_63433).jpg)

thick steel plate was a mere 21.2 m^2, which proved insufficient for constructing such large hulls. The incorporation of the 18-inch guns, known for their substantial weight, posed numerous design considerations. The investment required to increase weight capacity was deemed excessively expensive and was consequently abandoned. Furthermore, the firing pressure of these three 18-inch guns rendered close-range deployment impractical, thereby exposing a significant weakness in the anti-aircraft defense system. Ultimately, the failure of Yamato and Musashi serves as a stark example of the limitations of Japan's economic power and welding-related technology at the time.

6.5 STEEL IN MODERN WARFARE

Since the Iron Age, iron has been gradually replaced by steel as the material of choice for weapons. With the advent of gunpowder and firearms, the importance of steel, which could withstand high temperatures and pressures, grew significantly. Steel made it possible to increase the range of cannons and naval guns while reducing costs. The performance of crucial weapons such as tanks, warships, and submarines, which played decisive roles in warfare, relied heavily on the technology used in iron and steel manufacturing.

In modern warfare, with the increased prominence of missiles such as ICBMs (intercontinental ballistic missiles) and ABMs (anti-ballistic missiles), the direct impact of steel on weapon performance has diminished. However, the demand for high-performance steel products for precision weapon components remains consistent. While lightweight titanium and aluminum alloys are often employed for missile fuselages, special alloy steels are used for motor cases that house propellants due to their high strength and heat resistance. Submarines require high-strength steel with excellent weldability for streamlined hulls that enhance fuel efficiency, and there is also a need for non-magnetic steel to avoid detection by radio waves.

Starting from the Korean War, which escalated the East–West Cold War tensions, localized conflicts such as the Vietnam War, the Falklands War, the Gulf War, and more recently, the Iraq and Ukraine wars, have persisted. Throughout these armed conflicts, the involved parties have shown a keen interest in new weapons that improve upon the performance of existing ones, thereby driving the demand for advanced materials. This chapter provides a brief overview of special steels used in aerospace and for bulletproof purposes.

6.5.1 Special Steel for Aerospace

Steel materials used in the aerospace industry account for 10–20% of the total weight, but their price is often more than ten times higher than that of commercial steel. While steel has a density about three times higher than aluminum and 1.7 times higher than titanium, when considering specific strength, which takes into account both strength and specific gravity, high-strength steel can be comparable to high-strength aluminum and titanium alloys. As a result, high-strength steels with a yield strength of approximately 1,600 MPa have been developed for aerospace applications, and the potential for using ultra-high strength alloy steels is considered to be high due to their high-temperature characteristics, fracture toughness, and resistance to stress corrosion cracking.

Table 6.2 presents the composition and mechanical properties of advanced high-strength steels used in aerospace. Most of these steels exhibit a martensitic structure, and through appropriate heat treatment, it is possible to create transformation structures or fine precipitates.

TABLE 6.2 Composition and Mechanical Properties of High-strength Steel for Aerospace[73]

	Chemical Composition (wt. %)							Mechanical Property			
Alloy	C	Ni	Co	Cr	Mo	Mn	etc.	YS (MPa)	UTS (MPa)	K_{IC} (MPa$\sqrt{m}$)	K_{ISCC} (MPa$\sqrt{m}$)
4340	0.40	1.8	-	0.85	0.25	0.7	Si 0.2	1,482	1,965	71	11-16
300M	0.40	1.8	-	0.85	0.4	0.7	V 0.1, Si 1.6	1,689	1,965	71	11-16
HP9-4-30	0.30	9	4	1	1	0.2	V 0.108, Si 0.1 max	1,413	1,586	121	-
HY180	0.10	10	8	2	1	0.15		1,276	1,344	203	45
AF1410	0.16	10	14	2	1	-		1,551	1,689	187	45.71
15-5PH	0.04	4.6	-	15	-	0.25	Cu 3.3, Si 0.4	1,089	1,124	132	132
PH13-8	0.04	8	-	13	2.2	-	Al 1.1	1,434	1,551	81	>69
Maraging	0.005	16	8	-	2.2	-	Al 0.1, Ti 0.4	1,689	1,724	110	33

Maraging steel, located at the bottom of Table 6.2, is an ultra-low carbon Ni–Co–Mo alloy and represents the most recent development in ultra-high strength steel for aerospace.[73] It possesses extremely low carbon content, excellent weldability, and workability, making it a highly desirable material for various aerospace applications.

Maraging steel was initially developed in the US as a material for jets and missiles. The strength of steel can theoretically be increased to tens of thousands of MPa; however, this increase is accompanied by a deterioration in toughness and malleability. Maraging steel is categorized based on the amount of nickel added, and among the various grades, the most commonly used is the 18% Ni steel shown in Table 6.3.

Maraging steel contains a carbon content of less than 0.01%, making it relatively easy to weld and perform plastic work. The alloying elements added to different grades of maraging steel include nickel, cobalt, molybdenum, titanium, and aluminum. The term "maraging" is a combination of "martensite," which refers to a high-hardness microstructure, and "aging," which signifies the aging heat treatment process. Due to its low carbon content, the strength of maraging steel primarily stems from the martensitic matrix structure and the precipitates formed during aging.

The heat treatment process of maraging steel involves two steps: solution treatment and aging treatment. Solution treatment entails heating the material within the temperature range where austenite is stably formed. As depicted in Figure 6.30, the choice of aging temperature significantly affects the mechanical properties of the steel.[74] The resulting precipitates are in the form of Ni_3M, where M represents aluminum, titanium, and molybdenum, either individually or in combination. The strength of maraging steel largely depends on these precipitates, which, despite being very fine in size, have a volume ratio more than ten times greater than that of

TABLE 6.3 Types and Composition of Maraging Steel[74]

	Chemical Composition (wt. %)							
Alloy	**C**	**Ni**	**Co**	**Mo**	**Ti**	**Al**	**Mn**	**Si**
Grade 200	≤0.03	18	8.5	3.3	0.2	0.1	≤0.10	≤0.10
Grade 250	≤0.03	18	8	4.8	0.4	0.1	≤0.10	≤0.10
Grade 300	≤0.03	18.5	9	4.8	0.7	0.1	≤0.10	≤0.10
Grade 350	≤0.03	18	12	4.2	1.5	0.1	≤0.10	≤0.10
Grade 400	≤0.03	13	15	10	0.2		≤0.10	≤0.10
Grade 500	≤0.03	8	18	14	0.2		≤0.10	≤0.10

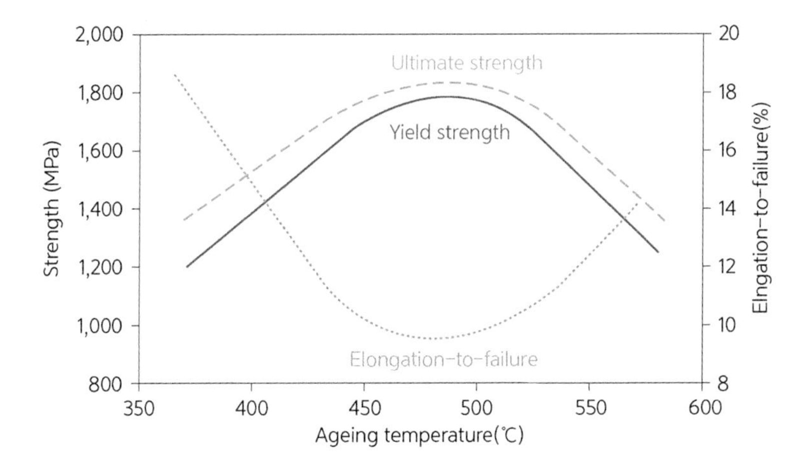

FIGURE 6.30 Effect of aging temperature on tensile properties of maraging steel.[74]

general high-strength steel. Furthermore, the precipitates form coherent interfaces with the matrix structure, enhancing the strengthening effect.

Close control of alloying is crucial in the manufacturing of maraging steel. The addition of cobalt raises the martensite formation temperature during cooling, while elements such as aluminum and molybdenum lower the martensite formation temperature. The martensite formation temperature directly influences the formation of the martensite structure, which is responsible for the high strength of the steel. Consequently, a higher martensite formation temperature makes it more challenging to achieve the desired martensitic structure and the associated high strength.

Moreover, the addition of cobalt reduces the solubility of molybdenum (Mo) in the base structure, facilitating the formation of precipitates. This promotes the uniform distribution of precipitates throughout the steel and contributes to a reduction in aging time.

Table 6.4 illustrates that maraging steel can achieve ultra-high strength within the range of 1,300 to 2,600 MPa. Its exceptional strength retention at high temperatures, around 500°C, makes it suitable for various applications such as rocket cases, jet engine parts, aircraft airframe components, ultra-high pressure chemical industry parts, and hot working dies and molds.

Although maraging steel is relatively expensive, its competitiveness is significantly higher when compared to aluminum or titanium-based alloys developed for aircraft or engine components where weight reduction is a top priority. Figure 6.31 showcases a motor engine, which serves

TABLE 6.4 Mechanical Properties of Maraging Steel[74]

Alloy	Aging temp. (°C)	YS (MPa)	UTS (MPa)	Fracture toughness, (MPa$\sqrt{m}$)	Charpy (J)
Grade 200	480	1,316	1,380	140	60
Grade 250	480	1,635	1,690	100	35
Grade 300	480	1,635	1,910	70	25
Grade 350	480	2,427	2,468	40	16
Grade 400	525	2,530	2,569	55	-

FIGURE 6.31 Motor engines of space rockets using maraging steel.[75] (Courtesy of https://ko.wikipedia.org/wiki/KRE-075#/media/%ED%8C%8C%EC%9D%BC:KARI_735kN_rocket_engine_for_KSLV-2.jpg)

as the propulsion system for missiles or space rockets, utilizing maraging steel.[75] Additionally, the use of maraging steel is expanding to include landing gear and take-off devices for helicopters and airplanes.

6.5.2 Armor Steel

To achieve victory in warfare, bulletproofing is essential as a defensive measure to minimize the damage from attacks and enhance defensive capabilities. Steel has been extensively employed as a material for bulletproofing, even with the advancement of gunpowder and bullets. However, recent efforts have focused on developing new materials and armor structure designs to further enhance armor performance.

The development of armor steel was driven by the need to counter weapons with increased firepower during World War II. One notable change in terms of materials has been the utilization of lightweight ceramics and fiber-reinforced plastics (FRP), as well as the adoption of multi-layer composite structures composed of two or more materials. Composite armor plates exhibit superior weight reduction compared to steel, but steel remains the primary choice for armor materials due to its economic feasibility, excellent workability, and high structural stability. Steel, as an armor material, enhances its protective ability with increasing thickness. However, thicker armor reduces mobility, so achieving lighter weight while maintaining excellent protection capability remains a significant challenge in modern times.

Bulletproof steel is categorized into alloys based on Mn–Mo, Cr–Mo, or Ni–Cr–Mo–(V). There are numerous medium-carbon and low-alloy steel, and the composition and type of alloying elements are determined to ensure adequate hardenability. Since welding is often involved in the manufacturing process, the carbon equivalent (CE) is typically kept below 0.1. The rolled product of bulletproof steel is known as rolled homogeneous armor (RHA) and undergoes heat treatment to achieve the necessary hardness, which is a critical mechanical property determining its bulletproof performance.[76]

Furthermore, high-hardness armor (HHA) steel, which exhibits superior performance compared to RHA, has been developed. HHA typically ranges from 477 to 534 HB (Brinell hardness), which is approximately 200 HB higher than RHA hardness. In the case of armored vehicles, the bottom part is optimized to withstand landmines, while the sides and front are designed to be bulletproof against small arms fire. As the hardenability of bulletproof steel determines its performance, it is essential to achieve a uniform martensite structure throughout the thickness of the material.

The failure mode of bulletproof steel varies depending on the impact conditions when the projectile strikes the bulletproof material. The key influencing factors include material properties, impact velocity, and projectile shape. Figure 6.32 illustrates the change in bulletproof modes according to the hardness of the bulletproof steel. Area A represents a mode where holes are formed in the bulletproof steel due to significant

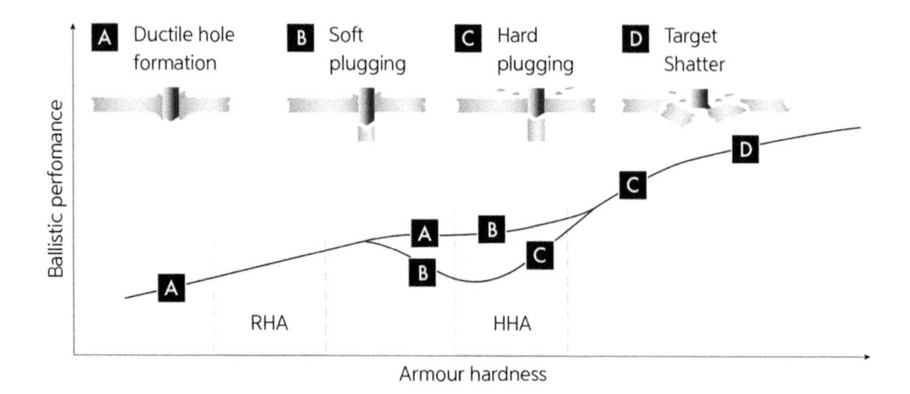

FIGURE 6.32 Hardness dependence of ballistic performance and failure mode of ballistic steel.[77]

plastic deformation. In areas A and B, the projectile's energy is absorbed through uniform plastic deformation of the bulletproof material. As hardness further increases, area C experiences fracture due to hard plugging. In this case, all of the projectile's energy is utilized to form locally observed Adiabatic Shear Bands (ASB) on the bulletproof material. In region D, characterized by very high hardness, the bulletproof material shatters due to its extremely low toughness. [77]

ASB (Adiabatic Shear Band) is a unique phenomenon observed in the high-speed deformation of bulletproof steel. If the thermal energy generated by the impact of bullets cannot dissipate effectively, the temperature in the shear band area rises rapidly, leading to a deterioration in bulletproof performance. The formation of ASB is influenced by several factors, including specific heat, heat transfer coefficient, specific gravity, and strain hardening coefficient. Decreasing the values of these factors promotes the formation of ASB, resulting in a localized increase in temperature, a decrease in shear stress, and an increased concentration of deformation, making the material more susceptible to failure.

In recent years, armor steels with new structures and compositions have been explored. One example is multi-layer steel plates, where a high-hardness front plate is combined with a high-ductility back plate. This configuration has shown significant improvements in ballistic performance. Manufacturing techniques such as carburizing, roll bonding, and explosive bonding are employed to produce multi-layer steel plates. Figure 6.33 provides a comparison of the ballistic capability between a dual hardness multi-layer steel plate, RHA, and HHA.[77]

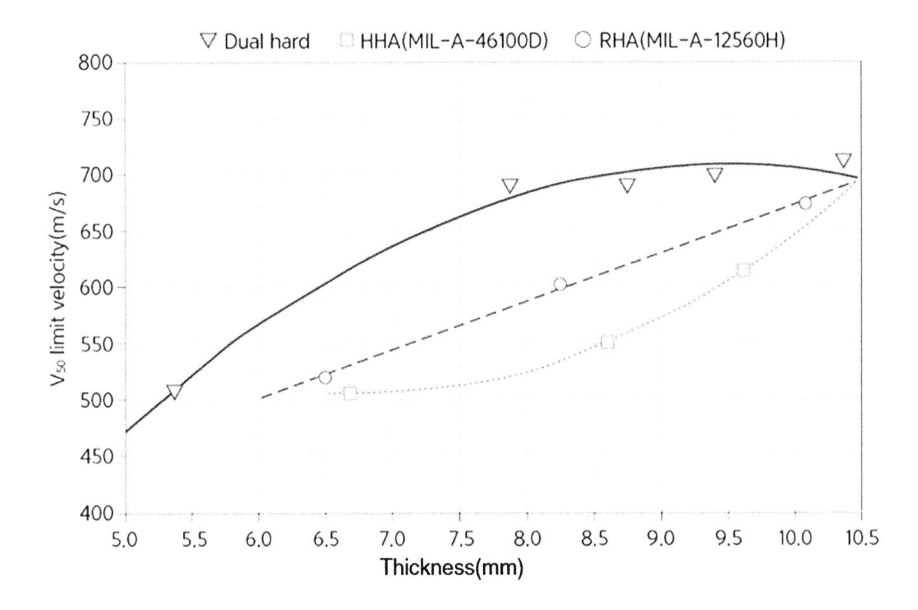

FIGURE 6.33 Comparison of bulletproof performance (V_{50}) of dual hardness bulletproof steel, RHA, and HHA.[76]

Maraging steel is also considered a potential material for armor due to its ability to form a high-hardness martensitic structure through slow cooling and its ease of plastic processing. Transformation Induced Plasticity (TRIP) steel, commonly used in the automotive industry for its high strength, and Twinning-Induced Plasticity (TWIP) steel show promise for bulletproof applications, and their development holds prospects.

REFERENCES

1. Chris Hedges, "What every person should know about war" July 6, 2003, https://www.nytimes.com/2003/07/06/books/chapters/what-every-person-should-know-about-war.html
2. Richard A. Gabriel and Karen S. Metz, "A short history of war: The evolution of warfare and weapons, strategic studies institute U.S. army war college", June 30, 1992, https://www.orbat85.nl/documents/DTIC/a255111.pdf
3. Christopher McFadden "7 Common but lesser-known siege weapons from antiquity", May 29, 2020, https://interestingengineering.com/innovation/7-common-but-lesser-known-siege-weapons-from-antiquity
4. "7 Powerful medieval weapons that characterized siege warfare", https://www.thecollector.com/siege-warfare-powerful-medieval-weapons/
5. "Berthold Schwarz", last edited December 10, 2023, https://en.wikipedia.org/wiki/Berthold_Schwarz

6. "Battle of Crécy", last edited March 9, 2024, https://en.wikipedia.org/wiki/Battle_of_Crécy
7. "Machine gun", March 9, 2024, https://www.britannica.com/technology/machine-gun
8. Nathaniel L. Erb-Satullo, Brian J. J. Gilmour, and Nana Khakhutaishvili, "The Metal Behind the Myths: Iron Metallurgy in Southeastern Black Sea Region", Cambridge University Press, 2019, https://www.google.co.kr/url?sa=t&rct=j&q=&esrc=s&source=web&cd=&cad=rja&uact=8&ved=2ahUKEwiIwruWkdr8AhWvsFYBHfsnBGsQFnoECDEQAQ&url=https%3A%2F%2Fora.ox.ac.uk%2Fobjects%2Fuuid%3A30186657-e0f1-412c-a1dd-1c0dc0c85dc3%2Ffiles%2Fm524554c68d3deb82a2edced7e3a62e5e&usg=AOvVaw1lU4Ou5PSZk1WALAiDhI5C
9. Gordon Doherty, "The Hittite destroyers: Chariot warfare in the late bronze age", https://www.gordondoherty.co.uk/writeblog/the-hittite-destroyers-chariot-warfare-in-the-late-bronze-age
10. Radomír Pleiner and Judith K. Bjorkman, "The Assyrian iron age: The history of iron in the Assyrian civilization", *Proc. Am. Philos. Soc.*, Vol. 118, No. 3, June 7, 1974, pp. 283–313, https://www.jstor.org/stable/986447
11. "Battle of Kadesh: Ancient Egypt vs The Hittite Empire", *The Collector*, https://www.thecollector.com/battle-of-kadesh/
12. Trevor Bryce, "The 'Eternal treaty' from the Hittite perspective", https://espace.library.uq.edu.au/view/UQ:263207/UQ263207_OA.pdf
13. "The Cyrus cylinder", *World History Encyclopedia*, https://www.worldhistory.org/article/166/the-cyrus-cylinder/
14. "Achaemenid empire", *World History Encyclopedia*, https://www.worldhistory.org/Achaemenid_Empire/
15. "Greece the Persian wars", http://www.historyshistories.com/greece-the-persian-wars.html
16. Jack Martin Balcer, "The Persian wars against Greece: A reassessment", *Historia: Zeitschrift für Alte Geschichte*, Vol. 38, No. 2, 2nd Qtr., 1989, pp. 127–143, Published by Franz Steiner Verlag, https://www.jstor.org/stable/4436101
17. "Gladius", last edited March 25, 2024, https://en.wikipedia.org/wiki/Gladius
18. E. A. Ginzel, "Steel in ancient Greece and Rome", 1995, http://dtrinkle.matse.illinois.edu/MatSE584/articles/steel_greece_rome/steel_in_ancient_greece_an.html
19. "Attila", last edited March 1, 2024, https://en.wikipedia.org/wiki/Attila
20. "Fall of the Western Roman empire", last edited March 26, 2024, https://en.wikipedia.org/wiki/Fall_of_the_Western_Roman_Empire
21. Dan Derby, "How the stirrup changed our world", September 24, 2001, http://strangehorizons.com/non-fiction/articles/how-the-stirrup-changed-our-world/
22. "The Hunnic war machine: Horsemen of the steppe – Part II", https://members.ancient-origins.net/articles/hunnic-war-machine-horsemen-steppe-%E2%80%93-part-ii

23. "Glimpse into the customs of the people of Goguryeo - The story of the murals of Muyongchong" (in Korean), https://wizztour.com/826
24. "Parthian shot", last edited February 5, 2024, https://en.wikipedia.org/wiki/Parthian_shot
25. "Battle of Carrhae", last edited March 30, 2024, https://en.wikipedia.org/wiki/Battle_of_Carrhae
26. "Migration age", https://www.worldhistory.org/Migration_Age/
27. "Crusades", last edited March 30, 2024, https://en.wikipedia.org/wiki/Crusades
28. John D. Verhoeven, "Genuine Damascus steel: A type of banded microstructure in hypereutectoid steels", *Steel Res.*, Vol. 73, No. 8, 2002, https://onlinelibrary.wiley.com/doi/10.1002/srin.200200221
29. "Damascus steel", last edited March 22, 2024, https://en.wikipedia.org/wiki/Damascus_steel
30. Gopi Katragadda, "The mystery of the Damascus sword and India's materials heritage", October 13, 2012, https://www.forbesindia.com/blog/technology/the-mystery-of-the-damascus-sword-and-indias-materials-heritage/
31. "Carbon nanotechnology in a 17th century Damascus sword", https://www.nationalgeographic.com/science/article/carbon-nanotechnology-in-an-17th-century-damascus-sword
32. M. Reibold, P. Paufler, A. Levin, et al. "Carbon nanotubes in an ancient Damascus sabre". *Nature*, Vol. 444, 2006, p. 286, https://doi.org/10.1038/444286a
33. Janet Lang, "Study of the metallography of some roman swords", *Britannia*, Vol. 19, 1988, pp. 199–216, https://www.jstor.org/stable/526199
34. "Excalibur", last edited March 21, 2024, https://en.wikipedia.org/wiki/Excalibur
35. "Gram (mythology)", last edited October 23, 2023, https://en.wikipedia.org/wiki/Gram_(mythology)
36. "Harpe", last edited January 13, 2024, https://en.wikipedia.org/wiki/Harpe
37. "Ascalon", https://highschooldxd.fandom.com/wiki/Ascalon
38. "Gladiator", last edited March 6, 2024, https://en.wikipedia.org/wiki/Gladiator
39. "Gan Jiang and Mo Ye", last edited May 7, 2023, https://en.wikipedia.org/wiki/Gan_Jiang_and_Mo_Ye
40. "Cao Cao's four great swords" (in Korean), https://m.blog.naver.com/PostView.naver?isHttpsRedirect=true&blogId=nightjaeho&logNo=70084468588
41. "Ssanggogeom" (in Chinese), https://baike.baidu.com/item/%E5%8F%8C%E8%82%A1%E5%89%91/3195837
42. "Kim Yushin" (in Korean), last edited March 9, 2024, https://ko.wikipedia.org/wiki/%EA%B9%80%EC%9C%A0%EC%8B%A0
43. "Chiljido" (in Korean), last edited March 20, 2024, https://ko.wikipedia.org/wiki/%EC%B9%A0%EC%A7%80%EB%8F%84
44. "Jeondo" (in Korean), https://namu.wiki/w/%EC%A0%84%EC%96%B4%EB%8F%84

45. “Ssangyonggeom” (in Korean), https://namu.wiki/w/%EC%8C%8D%EB%A3%A1%EA%B2%80
46. “Tenka-Goken”, last edited January 23, 2024, https://en.wikipedia.org/wiki/Tenka-Goken
47. “Great chain of the golden horn: Constantinople’s impenetrable barrier”, https://www.ancient-origins.net/ancient-places-asia/constantinoples-great-chain-0014317
48. “Mehmed II”, last edited March 27, 2024, https://en.wikipedia.org/wiki/Mehmed_II
49. George Anapniotis and Nikolaos Uzunoglu, “The great chain of the golden horn”, 2019, https://www.researchgate.net/publication/333804779_THE_GREAT_CHAIN_OF_THE_GOLDEN_HORN
50. “Dardanelles gun”, last edited December 8, 2023, https://en.wikipedia.org/wiki/Dardanelles_Gun
51. Roger Crowley, “The guns of Constantinople”, July 30, 2007, https://www.historynet.com/the-guns-of-constantinople/
52. “British empire”, last edited March 30, 2024, https://en.wikipedia.org/wiki/British_Empire
53. “Age of discovery”, last edited March 26, 2024, https://en.wikipedia.org/wiki/Age_of_Discovery
54. “The battle of Calais”, https://www.gyeongnam.go.kr/index.gyeong?menuCd=DOM_000010202005002000
55. “The Spanish Armada”, https://www.britishbattles.com/the-spanish-war/the-spanish-armada/
56. “Artillery through the Ages”, https://www.nps.gov/parkhistory/online_books/source/is3/is3c.htm%20
57. “Mary rose”, last edited March 25, 2024, https://en.wikipedia.org/wiki/Mary_Rose
58. “Walter Ralegh c.1552–1618, English explorer and courtier”, https://www.oxfordreference.com/display/10.1093/acref/9780191843730.001.0001/q-oro-ed5-00008718;jsessionid=8998ED0B2FF28D031DC786B66360385A
59. “Hammer and tongs”, https://www.liverpoolmuseums.org.uk/stories/hammer-and-tongs
60. “The first naval battle of the 21st century”, https://www.usni.org/magazines/naval-history-magazine/2022/february/first-naval-battle-21st-century
61. “Japanese cruiser Izumo”, last edited November 19, 2023, https://en.wikipedia.org/wiki/Japanese_cruiser_Izumo
62. “Theresa May inspects MSDF helicopter carrier at Yokosuka base”, https://www.japantimes.co.jp/news/2017/08/31/national/politics-diplomacy/may-inspects-msdf-helicopter-carrier-yokosuka-base/
63. “Pacific war”, last edited March 30, 2024, https://en.wikipedia.org/wiki/Pacific_War
64. “Forty-second annual report of the United States steel corporation”, https://digital.case.edu/islandora/object/ksl%3Auniann41
65. “Bethlehem steel, Hannah Hakim, the national world war II Museum”, https://salutetofreedom.org/pa.html

66. Arthur L. Herman, "Freedom's Forge: How American business produced victory in World War II", 2012, https://noahpinion.substack.com/p/book-review-freedoms-forge
67. "Lend-lease", last edited March 18, 2024, https://en.wikipedia.org/wiki/Lend-Lease
68. David C. Evans, "Planning pearl harbor", Thursday, April 30, 1998, https://www.hoover.org/research/planning-pearl-harbor
69. Yu Tianren, "Advisors in Daihonai, Nanam", 2014 (in Korean)
70. "10 Facts: Civil War Artillery", American Battlefield Trust, https://www.battlefields.org/learn/articles/10-facts-civil-war-artillery
71. Christopher J. Tassava, "The American economy during World War II", https://eh.net/encyclopedia/the-american-economy-during-world-war-ii/
72. Kitaro Matsumoto, "Design and construction of the Yamato and Musashi", *US Naval Institute Proceedings*, October 1953, Vol. 79/10/608, https://www.usni.org/magazines/proceedings/1953/october/design-and-construction-yamato-and-musashi
73. W. M. Garrison Jr., "Ultra-strength steels for aerospace applications", *JOM*, May 1990, https://www.researchgate.net/profile/Warren-Garrison/publication/257275543_Ultrahigh-strength_steels_for_aerospace_applications/links/57d19f2508ae0c0081e0508a/Ultrahigh-strength-steels-for-aerospace-applications.pdf
74. Adrian P. Mouritz (Editor), "Maraging Steel", Introduction to Aerospace Materials, Woodhead Publishing, 2012, pp. 15–38.
75. "KARI 75-ton rocket engine" (in Korean), last edited September 6, 2023, https://ko.wikipedia.org/wiki/KARI_75%ED%86%A4%EA%B8%89_%EB%A1%9C%EC%BC%93%EC%97%94%EC%A7%84#
76. W. Gooch, M. Burkins, D. Mackenzie, and S. Vodenicharov, "Ballistic analysis of Bulgarian electroslag remelted dual hard steel armor plate", 22nd Int. Symposium on Ballistics, Vancouver, BC, Canada, Vol. 2, November 14–18, 2005, pp. 709–716. https://www.researchgate.net/publication/292328380_Ballistic_Analysis_of_Bulgarian_Electroslag_Remelted_Dual_Hard_Steel_Armor_Plate
77. G. Crouch, S. J. Cimpoeru, H. Li, and D. Shanmugam, "The Science of Armour Materials", Woodhead Publishing in Materials, 2017, pp. 55–115, https://doi.org/10.1016/B978-0-08-100704-4.00002-5

CHAPTER 7

Steel Industry, Today

7.1 THE RISE AND FALL OF THE US STEEL INDUSTRY

In 1850, the amount of iron and steel production in the US was only one-fifth of that in Britain. After the Civil War, however, American business-people became interested in the innovative productivity of the Bessemer converter. The US also had a high demand for infrastructure construction, including road paving, city connectivity, bridge construction over rivers, and the development of railroads throughout the West. To meet this growing demand, US steelmakers adopted the blast furnace for iron-making and utilized the Bessemer converter and open-hearth furnace for steelmaking. This led to the continuous expansion of production capacity. With the introduction of these innovative technologies and facilities, the US was able to minimize the need for trial and error.

Steel tycoon Andrew Carnegie (1835–1919) depicted in Figure 7.1, is a symbol of the "American Dream." At the age of 12, he settled in a poor neighborhood in Pittsburgh, Pennsylvania. While working as a telegram carrier and later as an employee in a railroad company, Carnegie gained valuable experience and a broader business perspective. He strategically invested in stocks of bridge construction companies, railroad factories, locomotive factories, and steel mills. After the end of the Civil War in 1865, the 30-year-old Carnegie developed a keen interest in bridge construction and owned a steel mill that produced the necessary materials. Initially, his steel mill primarily produced cast iron. However, recognizing the superior mechanical properties of steel for building durable bridges, Carnegie adopted the Bessemer converter in his steel mill. The steelworks

DOI: 10.1201/9781003419259-7

FIGURE 7.1 Steel King Andrew Carnegie.[1] (Courtesy of https://billofrightsinstitute.org/essays/andrew-carnegie-and-the-creation-of-us-steel)

shown in Figure 7.2 were constructed in Pittsburgh, Pennsylvania. Carnegie's steel was not only used for bridges but also played a vital role in the emergence of a new architectural marvel known as the "skyscraper." In 1889, Carnegie consolidated all of his businesses into a single corporation, Carnegie Steel.[1]

Following the success of Carnegie's steel company, numerous other steel and mining companies were established throughout the US. American steel production skyrocketed from 220,000 tons in 1873 to 11.4 million tons in 1900, surpassing the combined production of Britain and Germany.[2] US Steel (USS) was founded in 1901 by E. H. Gary, in collaboration with Carnegie, J. P. Morgan, and C. Schwab.[1] The USS became the world's largest company and accounted for two-thirds of all steel production in the US. In 1904, C. Schwab founded BSC.[3] BSC quickly grew to become the second-largest steel company in the US, second only to USS, and particularly excelled in the shipbuilding sector. As we saw in Chapter 6, BSC was able to thrive during World War I and World War II by fulfilling substantial orders for military equipment, including naval vessels, from allied countries, as shown in Figure 7.3.

With the outbreak of World War I in 1914, American steel production experienced further growth. By 1913, steel production in the US

FIGURE 7.2 Pittsburgh Steel Complex (1906). (Courtesy of https://en.wikipedia.org/wiki/File:Mills_in_Strip_District,_Pittsburgh_(84.41.70).jpg)

FIGURE 7.3 Bethlehem Steel workers made artillery shells for World War II. (Courtesy of https://salutetofreedom.org/pa.html)

had exceeded 30 million tons, nearly three times the amount produced in 1900.[4] In the 1920s and 1930s, the skylines of New York and Chicago saw the rise of Art Deco–style skyscrapers. BSC's steel played a significant role in the construction of landmarks such as Rockefeller Center, the Waldorf Astoria Hotel, the George Washington Bridge, and the Golden Gate Bridge.[5] In 1930, BSC's steel was utilized in the construction of the Chrysler Building, which, at the time, held the title of the world's tallest building. Less than a year later, the Empire State Building, using around 60,000 tons of USS, claimed the title of the tallest building.[6]

Innovations in the steel industry also contributed greatly to the mass production of automobiles, home appliances, and food cans. The asset values of USS and BSC surpassed those of Ford and General Motors. Although the Great Depression led to a slowdown in steel production along with the stock market crash, railroads continued to be constructed across the country, canned food remained popular, and the end of Prohibition saw the emergence of new steel products for beer cans. With the conclusion of the Great Depression and World War II, the US assumed full responsibility for supplying military equipment to the Allied Powers, securing its position as the world's largest and most competitive steel producer.[4, 7]

During World War II, the steel mills in the US operated around the clock, as twenty-four hours a day was not enough to meet the demand. Female labor was also mobilized to support the war effort. The economy began to thrive once again, and American steel production quickly surpassed that of any other country, reaching over three times the output of its closest competitor. Throughout the war, the US produced 25 times more steel than it did during World War I, and American steel products played a crucial role in securing victory for the Allies. Following the war, the markets for automobiles, home appliances, toys, and construction materials experienced unprecedented growth. Scrap from dismantled ships and tanks was recycled to produce bridges and beer cans. The steel demand continued to rise, reaching 130 million tons in 1973.[7]

However, the heyday of American steel did not last long. The emergence of the United Steel Workers of America in the 1950s led to frequent strikes, and wage increases followed the two oil shocks. Additionally, logistical inefficiencies in inland steelworks contributed to the decline of the US steel industry. After World War II, the US steel industry, which once held a dominant 64% share of the global market, saw its market share decline to 20% in the 1970s and further drop to 10% in the 1980s, as illustrated in Figure 7.4. The increasing cost burden hindered the adoption

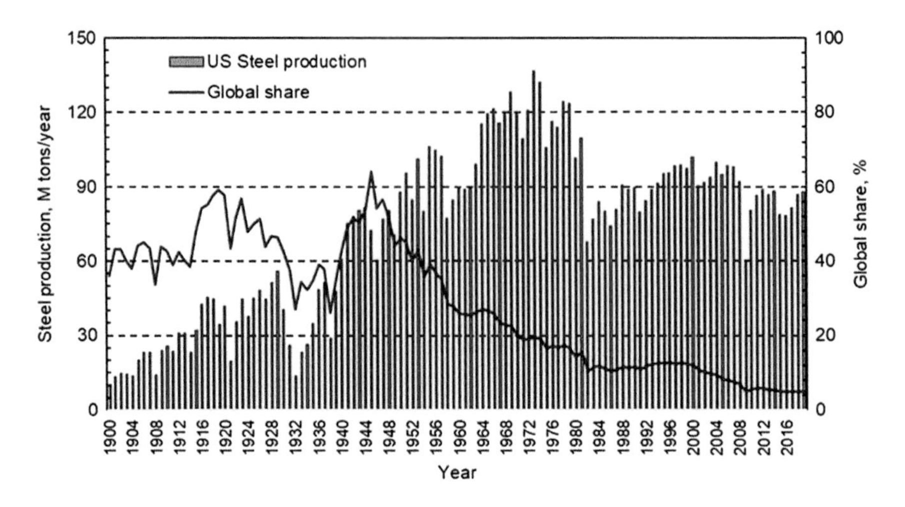

FIGURE 7.4 Changes in US steel production and global share.

of emerging technologies such as LD converters and continuous casting when needed. The US steel industry underwent restructuring, including the merger of USS and BSC, but improving competitiveness proved challenging. Eventually, Japan surpassed the US in steel production.(7)

However, the steel industry in the US did not experience a complete decline. The US benefits from having the lowest electricity prices and an abundant supply of scrap generated from a long history of steelmaking. As a result, electric arc furnace (EAF) operations are more advantageous in the US compared to other countries. This advantage has led to the emergence of mini-mill companies that utilize EAF technology.

One prominent example is Nucor, which originated as an automobile company in 1906 and transformed into a steel company in the 1960s with a focus on the mini-mill concept. Currently, Nucor has surpassed USS and become the top steelmaker in the US, with its steel products accounting for over 50% of the total US production. Furthermore, as the operation of blast furnaces is recognized as a major contributor to CO_2 emissions and climate change, the higher proportion of EAF operations in the US steel industry may serve as a competitive advantage in the future.(7)

7.2 THE SUCCESS AND STAGNATION OF THE JAPANESE STEEL INDUSTRY

Until the end of the 19th century, Japan primarily relied on small-scale production systems for household items and weaponry. Japan adopted the Tatara and reverberatory furnace processes from Europe. Observing the Qing Dynasty's defeat at the hands of European powers during the Opium War, Japan recognized that mass-producing steel for military equipment was crucial for becoming a powerful nation. It was determined that the Tatara process would be insufficient to supply the large quantities of steel needed for various weapon manufacturing, leading Takato Oshima (大島高任) to construct Japan's first blast furnace in Kamaishi, Iwate Prefecture.[8]

In 1887, as part of the national policy of "Rich Nation, Strong Army" following the Meiji Restoration, Tanaka Ironworks operated a blast furnace based on European models. By 1893 (Meiji 26), approximately 8,000 tons of pig iron were produced in the blast furnace, accounting for more than half of the nationwide production of 14,654 tons.[8, 9] Following the victory in the Sino–Japanese War, Japan invested war preparations into establishing the national Yawata Steelworks in Kyushu in 1901. During World War II, numerous steelworks were constructed to meet the demand for steel in munitions production, and by the mid-1940s, crude steel production had surpassed 3.5 million tons.[8]

Japanese steelmakers' early adoption of LD converters serves as a prime example of their exceptional initiative and judgment regarding innovative technologies. The LD converter significantly reduced steelmaking time compared to the open-hearth furnace, requiring only a fraction of the operators. In 1976, four and a half years after its initial implementation, Japanese steelmakers transitioned their open-hearth furnaces to 100% LD converters. Concurrently, they made comprehensive improvements to facilities and operations to further enhance their competitiveness. Notable technological advancements included the introduction of a multi-hole lance and an oxygen converter gas recovery facility, increasing the converter's heat capacity to over 150 tons. The steelmaking time for a single heat was reduced to less than 30 minutes. Additionally, CC technology was first implemented in 1955 and achieved near perfection by 1990.[10]

In the late 1950s, the Japanese steel industry experienced a resurgence by establishing a domestic support system for policies and funds, receiving technological support from the US, and importing raw materials through international cooperation. It experienced rapid growth during the 1960s and, following the oil shock in 1973, became the world's largest producer and the most competitive, as depicted in Figure 7.5. Several factors contributed to its competitiveness, including the decline of the US steel industry, demand for post-war recovery and the Korean War, the

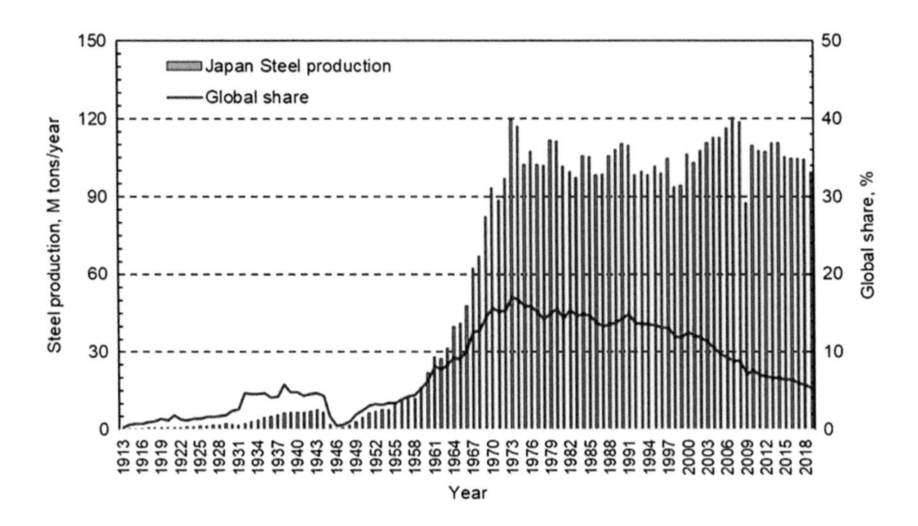

FIGURE 7.5 Changes in Japanese steel production and global share.

active promotion of technology development, and the establishment of an efficient production system with streamlined logistics.

Steelworks were strategically established along the coast, enabling a quick and efficient supply of massive quantities of raw materials. The concept of an integrated iron and steelmaking production system, where the entire process from raw material unloading to product shipment is interconnected, was first pioneered by Yataro Nishiyama, the president of Kawasaki Steel, and later adopted by all newly constructed Japanese steelworks. Steelworks built based on this concept offer logistical and cost advantages, especially when a significant portion of the raw materials is imported from abroad. Therefore, Korea's POSCO, China's Zhanjiang, and Fangchenggang Steelworks, which have followed this concept, have achieved world-class cost competitiveness [10]

Following World War II, Japan's annual steel production dropped to 560,000 tons in 1946. However, thanks to the successful implementation of various innovative initiatives mentioned above, production gradually increased to 4.84 million tons in 1950, 22.14 million tons in 1960, and 120 million tons in 1973. In 1973, Japan accounted for 17.1% of global steel production and became the world's largest exporter, capturing 28.6% of the global export market, as illustrated in Figure 7.5. Nevertheless, Japan's strong competitiveness entered a period of stagnation due to external factors such as the two oil shocks in the 1970s and the strengthening of the yen resulting from the Plaza Agreement in 1985.[11, 12]

The global economic downturn caused by the oil shock resulted in a significant decrease in steel demand, particularly impacting steelmakers with high debt ratios as a consequence of extensive facility investments. Japan's crude steel production, which reached 120 million tons in 1973, fell to 100 million tons just 2 years later. Furthermore, the Plaza Agreement in 1985 further impacted the Japanese steel industry. The value of the yen, which was at 250 yen per dollar in 1985, rose to 160 yen in 1986 and 79 yen in 1995. The strong yen significantly reduced the export competitiveness of the Japanese steel industry, leading to a crisis. In response, the industry made vigorous efforts to overcome this situation, including technology development for cost reduction, expanding domestic demand, and aggressively promoting exports despite facing accusations of dumping.

Technological advancements have yielded remarkable results in cost reduction and quality improvement within the Japanese steel industry. Blast furnaces, which constitute a significant portion of the production cost, have increased in size and longevity. In the 1970s, their lifespan was only 5 to 7 years, but by the 2000s, it exceeded 20 years. As of 2013, out of the 27 active blast furnaces, 20 had a net volume of 4,000 m^3 or more, and 13 super-large blast furnaces with a volume of 5,000 m^3 or more were constructed. Emphasis was placed on utilizing low-cost raw materials, developing energy-saving technologies like hot direct rolling, and enhancing process efficiency through integration and automation.[(10)]

Japanese steelmakers continuously responded to customer demands by developing new steel products with superior strength, machinability, weldability, and toughness. Despite the prevailing economic downturn, they made consistent investments in surface treatment facilities such as GA, which offers excellent weldability and workability. During this period, one-third of the forty-nine GA facilities established in 2002 were built. In the high-profit automotive steel sector, the industry implemented a guest engineer system where steel company engineers were dispatched to collaborate with customers in the development of new products. This approach facilitated reduced trial and error periods through close cooperation with customers from the product development stage, effectively deterring latecomers from easily catching up.[(10)]

During challenging times, the Japanese steel industry demonstrated astuteness in overcoming crises by achieving significant technological advancements in both operations and products. They also established an efficient cooperative system among the government, businesses, and society to achieve common goals. These aspects serve as valuable lessons for latecomers who seek to thrive in the industry.

7.3 THE GLOBAL STEEL INDUSTRY: THE AGE OF HYPER-COMPETITION

7.3.1 Steel Industry in Korea: Born from Scratch

The steel industry in Korea had been very weak until the establishment of POSCO. Just before World War II, the total crude steel production capacity of both South and North Korea was 350,000 tons per year, with most of the factories located in North Korea and very little production in South Korea. In 1965, South Korea's crude steel production was a mere 270,000 tons. However, in 1973, with the completion of POSCO's integrated steelworks, which included facilities for ironmaking, steelmaking, hot rolling, and plate rolling, the country's crude steel production capacity reached 1 million tons.[13, 14]

POSCO, founded in 1968, operates two steelworks in Pohang and Gwangyang, both strategically located on the coast. The company adopted an integrated production system. During the late 1960s, when the national income per capita was less than US$200, and with limited capital, technology, experience, and resources, there were many challenges and oppositions to constructing large-scale integrated steelworks capable of producing over 1 million tons per year. Despite these difficulties, POSCO has since become the most competitive steel company, with the capacity to produce about 40 million tons per year across its two steelworks in Korea.[13] Hyundai Steel, which initially produced steel solely through EAF, later constructed a large-capacity BF in Dangjin in 2010. Subsequently, two more BFs were added to operate an integrated production system.[15]

At the time of POSCO's establishment, President Park Chung-hee provided full support, recognizing the urgent need for steel in the construction of roads, ports, factories, and housing, which were the foundations of Korea's economic growth. Despite numerous difficulties during the construction and initial operation process, POSCO's first chairman, Park Taejoon, displayed unwavering determination, excellent management skills, a strong sense of mission, and the dedication of all employees. As a result, the company achieved normal operation in a short period and recorded a surplus from its first year of operation.[13]

Commencing with the completion of the first stage in Pohang in 1973, with a capacity of 1.03 million tons, POSCO progressively expanded the Pohang steel mill in several stages, reaching an annual capacity of 8.5

million tons by 1981. Simultaneously, the construction of the world's largest steelworks in Gwangyang commenced. With a production volume of 37.21 million tons, POSCO has established itself as one of the world's top five steel companies. Furthermore, according to World Steel Dynamics (WSD), a renowned global analytics, research, and consulting firm, POSCO has consistently been ranked as the world's most competitive steel company for the past 13 years, as depicted in Table 7.1.[16]

In 2021, Korea's steel production exceeded 70 million tons, ranking fifth or sixth globally in terms of production volume. POSCO, with its eight BFs and two FINEX facilities, plays a significant role in the industry. Hyundai Steel, with a yearly crude steel production capacity of 12 million tons, operates three BFs along with multiple EAFs. Furthermore, several EAF-based companies contribute to the production of construction and specialized steel. The growth of the Korean steel industry over the past 60 years is regarded as a model case laying the foundation for the growth of the national economy and national income, as shown in Figure 7.6.

In recent years, Korean steel companies have focused on developing high-value-added products that prioritize quality. Notably, they have concentrated on producing steel for automobiles, shipbuilding, and energy sectors. Korean steelmakers face a unique challenge as they find themselves sandwiched between Chinese competitors, who excel in mass production, and Japanese steelmakers, who specialize in high-end steel products. To maintain their competitiveness in terms of price, quality, and performance, Korean steel companies are actively pursuing intensive technology development.

7.3.2 China, Pushing with Volume

China's steel production underwent remarkable growth driven by the rapid expansion of its economy. In 1995, China produced approximately 100 million tons of steel, but by 2018, it surpassed 900 million tons, securing the top position globally and accounting for over half of the world's total steel production, as depicted in Figure 7.7.

China's focus on steel production capacity was primarily evident through the expansion efforts of national steel companies, supported by the central government. One notable example is the Baowu Steel Group, which constructed large-scale steelworks such as Zhanjiang and Fangchenggang, situated along the coast and featuring integrated production systems. Additionally, there are approximately 400 small- and medium-sized steel mills operating inland.

TABLE 7.1 Overall Competitiveness Ranking of Steelmakers Since 2010 Announced by WSD

Year/rank	1	2	3	4	5	6	7	8	9	10
2010	POSCO	JSW	Nucor	SAIL	CSN	NLMK	Tata/Corus	Usinimas	Severstal	Gerdau
2011	POSCO	Nucor	NLMK	Severstal	A. Mittal	NSC	JSW	CSN	SAIL	JFE
2012	POSCO	NLMK	CSN	Severstal	Bao	JSW	SAIL	NSC	Nucor	Sumitomo
2013	POSCO	Severstal	Nucor	NLMK	JSW	Gerdau	NSSMC	SAIL	Jindal	JFE
2014	POSCO	Nucor	NSSMC	Gerdau	Severstal	NLMK	JSW	JFE	Hyundai	Endermir
2015	POSCO	Nucor	NSSMC	Gerdau	Severstal	JSW	NLMK	JFE	Hyundai	Endermir
2016	POSCO	NSSMC	Nucor	SDI	NLMK	Severstal	VoestAlpine	Gerdau	JFE	JSW
2017	POSCO	Severstal	Nucor	NLMK	NSSMC	JSW	JFE	A. Mittal	Voest Alpine	Baowu
2018	POSCO	Nucor	Voest Alpine	Severstal	NSSMC	NLMK	A. Mittal	JSW	JFE	MMK
2019	POSCO	Nucor	Voest Alpine	Severstal	NSC	NLMK	JSW	A. Mittal	Evraz	Hyundai
2020	POSCO	Nucor	Severstal	NLMK	Voest Alpine	NSC	A. Mittal	SDI	JSW	Evraz
2021	POSCO									

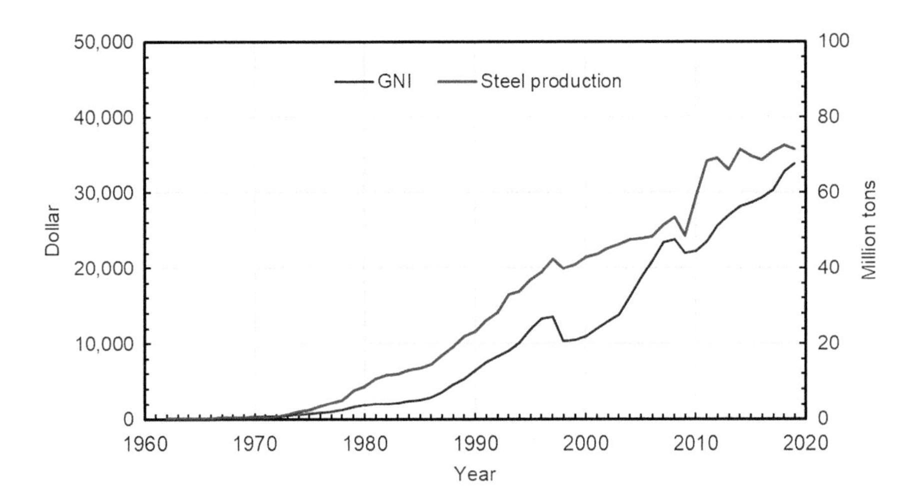

FIGURE 7.6 Changes in Korea's steel production and GNI from 1960–2020.

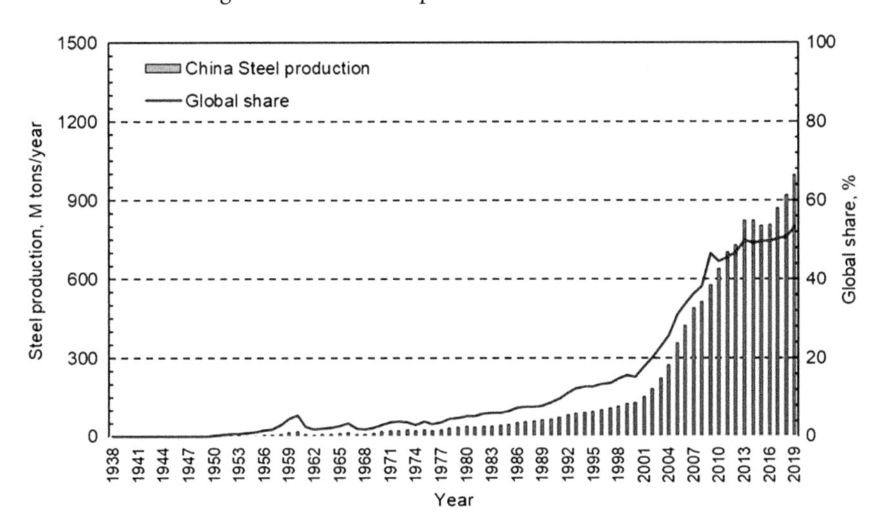

FIGURE 7.7 Changes in China's steel production and global share.

Many of these smaller steelmakers were established based on China's engineering capabilities. They are strategically located near raw material suppliers and customer bases. These steelmakers utilize various facilities such as EAFs, small BFs, high frequency induction furnaces, and cupolas to produce common steel products. Recent government measures have led to the closure of numerous small companies that primarily produced lower-quality Titao steel (地條鋼).

On the other hand, large-scale steel companies located in coastal areas predominantly import raw materials from foreign countries and export their production. The Chinese government, under the 2005 Steel Industry Development Policy, has not only encouraged the establishment of such large-scale steel companies but also promoted their expansion through mergers between existing major steel companies. Noteworthy examples include Baowu, which strives for the top two positions in global steel production through mergers and acquisitions (M&A), as well as Hebei, Sagang, Ansan, Shougang, and Shandong steel companies.[17, 18]

China emerged as the dominant power in Asia during the Iron Age, thanks to its pioneering smelting techniques and cast steel technology, which were first introduced during the Han Dynasty BC. However, this early advancement did not lead to the development of innovative technologies for mass steel production.[19–23] As a result, many of the current processes and technologies used in the Chinese steel industry have been recently imported from Europe and Japan.

In China, the steel industry primarily thrived in regions rich in iron ore and coal located inland. The development of related industries was closely tied to these local conditions. Consequently, even after the year 2000, outdated facilities that were in line with the scale of local logistics and the economy continued to play a significant role in steel production. This stands in contrast to Korea's POSCO, which constructed state-of-the-art facilities and fostered an environment of exceptional competitiveness during the 1970s and 1980s, ahead of China. In recent years, China has made substantial efforts to introduce cutting-edge processes and technologies from abroad, resulting in the construction of numerous modern and high-end steelworks. Examples of these advancements include the implementation of modern facilities such as LD steelmaking, CC, RH degasser, thin slab casting, CAL, and Continuous Galvanizing Line.

The uneasy relationship between China and Japan transitioned into a cooperative mode following the establishment of diplomatic ties in 1972. One notable example is the collaboration between NSC (Nippon Steel Corporation) and Wuhan Steel in the production of electrical steel sheets. NSC and JFE also cooperated with Bao (now Baowu Steel), which introduced the concept of large-scale steelworks located in coastal areas for the first time in China.[353] This cooperation went beyond simple technical support and included construction funds. Additionally, collaboration with Japanese steel companies extended to the product field. Baowu partnered with NSC and AM to establish a high-tech hot-dip galvanizing company called BNA, catering to the demand for automotive steel sheets from Japanese and European automobile companies operating in China.[24]

In the 2000s, the Chinese steel industry experienced rapid growth and began adopting governance structures akin to those of advanced countries. Through government policies, the number of steelmakers, which once exceeded 400, was halved, and two steel companies within the top three globally in terms of production scale were formed through mergers and acquisitions. Simultaneously, efforts were made to restructure the industry, including the closure of excess facilities, and significant investments were made in environmental initiatives. As the steel-consuming industries progressed, there was a notable increase in demand for high-value-added products, accompanied by attempts at distribution innovation through e-commerce. This trend is expected to continue until 2025, and future developments in the industry are a subject of great interest.(18)

7.3.3 Steel Industry in India and ASEAN

India has expanded its steel production capacity to match the growing economy. In 2018, its crude steel production exceeded 100 million tons, making it the world's second-largest producer. Major steelmakers, such as state-owned SAIL and RINL, as well as private companies such as JSW, Tata, Essar, and JSPL, are responding to the growing demand for steel. Tata has a strategic alliance with Japan's NSC,(25) JSW is cooperating with Japan's JFE by participating in a 15% stake,(26) and SAIL is cooperating with POSCO in the technological field of upstream steelworks.(27) The Indian government is pushing to advance the steel industry by restructuring Essar and selling it to an AM-NSC joint venture.(28)

The ASEAN region is a steel import market. High-grade steel has traditionally been imported from Japan, and most commercial-grade steel has been imported from China. Even in the 2000s, no country had a large-scale BF production system, and only steelmakers focused on EAFs were operating. In 2013, Indonesia started the operation of PT-KP, owned jointly with POSCO, an integrated steelworks equipped with ironmaking, steelmaking, and plate rolling facilities. The construction and operation of PT-KP were supervised by POSCO.(29) In 2018, Formosa, a Taiwanese company, built steelworks on a coastal area with two large BFs and an annual production capacity of 7 million tons, providing an opportunity to improve competitiveness in Vietnam.(30) ASEAN governments are emphasizing the importance of investment in the steel industry, but it is not easy to expand additional facilities due to high investment costs and stagnant demand.

7.3.4 European Steelmakers in Endless Transformation

Europe, with its long history of steel, is the birthplace of the modern steel industry. Britain, once the pioneer of the Industrial Revolution and responsible for nearly half of the world's steel production in the mid-19th century, lost its manufacturing competitiveness as its economic structure shifted towards the service industry. The state-owned British Steel Co. transformed a merger with Hoogovens of the Netherlands, becoming Corus. However, it still struggled to improve profitability and was eventually sold to India's Tata Group in 2007. It is now operated as Tata Steel Europe (TSE).[31]

In contrast to the UK, Germany developed an economic structure centered around manufacturing rather than the service industry. This led to an increased demand for steel products, expanded markets, and upgraded production technologies to maintain high competitiveness. As a result, steel production in Germany significantly increased throughout the 20th century, reaching 19 million tons in 1918 and 40 million tons in 1970. After the 1980s, Germany also experienced a recession, leading many steelmakers to undergo mergers and acquisitions for restructuring. Currently, Thyssen Krupp Steel (TKS) and Salzgitter operate as independent integrated steelworks, along with small rolling or forging mills and special steel companies based on EAFs. Thyssen and Krupp, founded in 1810 and 1891 respectively, merged in 1999 to create TKS. TKS had a reputation as a steelmaker with high technological competitiveness. However, due to deteriorating business conditions, discussions of a merger with a competitive steelmaker in the region have been taking place to improve competitiveness.[32]

AM, a multinational corporation, was formed in 2006 through the merger of Usinor-Sacilor, a French state-owned steel group, with steelmakers in Belgium, Spain, and Luxembourg. Subsequently, it merged with several steel companies in Italy, the CIS, South Africa, Argentina, Brazil, and the US, becoming the world's largest steelmaker with an annual production capacity of nearly 100 million tons.[33] Mergers and acquisitions are still ongoing at AM. In 2014, it acquired the Calvert steel company in Alabama, US, from TKS and operates it in a joint venture with NSC.[34] Recently, AM gained control of the Ilva steelworks in Italy[35] and continues to explore opportunities to expand its steel business in India and China. AM is recognized for its competitiveness in the steel business, commensurate with the size of its operations, and it strengthens its competitiveness by continuously developing new processes and technologies through collaboration with the company's internal technical staff and leading European engineering firms. In particular, AM focuses its efforts on the development and sale of unique products, concentrating on high-margin steel sheets for automobiles.

7.3.5 Securing the Competitiveness of the Steel Industry

The global steel industry has recently undergone significant changes in all aspects, including management, technology, markets, and the environment, and these changes will continue in the future. To survive in this situation, it is crucial to secure competitiveness. While the steel manufacturing process has a history of 4,000 years, major innovations, such as the BF and Bessemer converter, have been developed since the 18th century. Over the past 100 years, important modern core technologies, such as integrated and continuous processes, have also been developed.

During the last 50 years, numerous new steelworks have been constructed worldwide. The state-of-the-art steel facilities and operational technology currently in use are well-positioned to achieve standardization in a short period, unlike in the past. Equipment supply is handled by engineering companies with nearly a century of experience in this field, and trial and error can be reduced because operational technology can be transferred from existing steelmakers with extensive operational experience. Therefore, in terms of commercial steel production, it is considered that there is not a significant difference in competitiveness between countries or steel companies.

If a new steel company becomes competitive in a short period, it may create competition for existing steel companies. However, on a global scale, it may be more beneficial to divide sales territories and establish collaborative partnerships within the same industry. Advanced steelmakers in Europe and Japan are selectively choosing companies from a strategic perspective of globalization, exchanging strong technologies with each other, and expanding their material export areas through this approach. However, when it comes to high-value-added products, the situation is different. These products reflect unique and stringent customer requirements, and their production involves specialized operational know-how. Since this know-how is closely guarded, it is challenging for competitors to easily replicate it.

From the customer's perspective, optimizing and standardizing processes such as forming, cutting, welding, and surface treatment that are involved in manufacturing products using purchased steel materials is crucial. Therefore, customers are often reluctant to change their steel supplier because any alteration in the properties of the steel material due to a change in the source can result in deviation from the optimal range and lead to decreased productivity. This tendency is especially pronounced in the case of steel sheets for automobiles, where numerous manufacturing lines employ robotic systems. Therefore, when automakers expand their business to other countries, they strive to maintain the conditions for automobile production from their home countries by bringing along multiple materials and parts suppliers.

In recent times, steelmakers have been making efforts to enhance the added value of their customers' products through unique materials, technologies, processes, and components that are not readily available from other companies. This activity is aimed at fostering long-term relationships with customers. As part of these efforts, advanced steel companies establish long-term marketing systems by implementing a "Resident Engineer" or "Guest Engineer" program, in which internal engineers are carefully selected and assigned to work closely with customers.(10) From this perspective, the competitiveness of steel companies in the future may not solely depend on their operations or production capabilities, but also on their intangible assets such as product innovation and marketing capabilities.

It can be argued that the current steel industry has entered an era of intense competition, different from the past. With overcapacity and sluggish demand, competition among steelmakers has intensified, fueled by constant M&As. Consequently, companies need to find ways to supply higher-quality products at lower prices than their competitors to maintain their competitiveness. In the future, marketing strategies that generate added value through the development of superior products and optimization of customer processing procedures are expected to provide an advantage over tangible factors like product quality alone.

7.4 SHADES AND CHALLENGES OF THE STEEL INDUSTRY

The steel industry has achieved continuous quantitative growth since 1970, producing about 1.9 billion tons of crude steel worldwide in 2019, as shown in Table 7.2. Steel is the third most abundant man-made bulk material on Earth, after cement and timber.(36) Iron and steel production is a highly energy-intensive industrial activity, with the sector accounting for 20% of industrial final energy consumption and around 8% of total final energy consumption. The steel sector is also a large contributor to

TABLE 7.2 Top Ten Steel Producers and Estimated Market Share in 2019[36]

Rank	Company	Steel production, Mt	Global share, %
1	ArcelorMittal	97.3	5.2
2	Baowu	95.5	5.1
3	NSC	51.7	2.8
4	HBIS	46.6	2.5
5	POSCO	43.1	2.3
6	Shagang	41.1	2.2
7	Ansteel	39.2	2.1
8	Jianlong	31.2	1.7
9	Tata	30.2	1.6
10	Shougang	29.3	1.6
Top 10		**505**	**27**
Total		**1,869**	**100**

the current challenge we face in meeting our climate goals: direct CO_2 emissions from the sector are around 2.6 Giga tons of carbon dioxide (Gt CO_2) per year, or around a quarter of industrial CO_2 emissions, owing to its large dependence on coal and coke as fuels and reduction agents. This is equivalent to about 7% of total emissions from the energy system, including industrial process emissions.[36]

Most of the top ten steelmakers in Table 7-2, which account for 27% of global steel production, have declared carbon net zero by 2050 and are taking active measures to mitigate CO_2 emissions. The countermeasures can be broadly divided into the following categories.

1. Reduced coke rate in the blast furnace, increased scrap metal usage in the converter, and reduced energy consumption under the current operating system
2. Shifting from blast furnace production to hydrogen reduction steelmaking and electric furnace construction.
3. Utilization of emitted CO_2 by Carbon Capture and Storage (CCS) or Carbon Capture, Utilization and Storage (CCUS)

Even if the countermeasures in 1) are applied to integrated steelworks operating large-volume blast furnaces, it is expected that there will be limits to CO_2 emission reduction. Therefore, measures 2) and 3) are expected to become mainstream in the future. However, to commercialize these measures, in addition to the problem of how to obtain the enormous amount

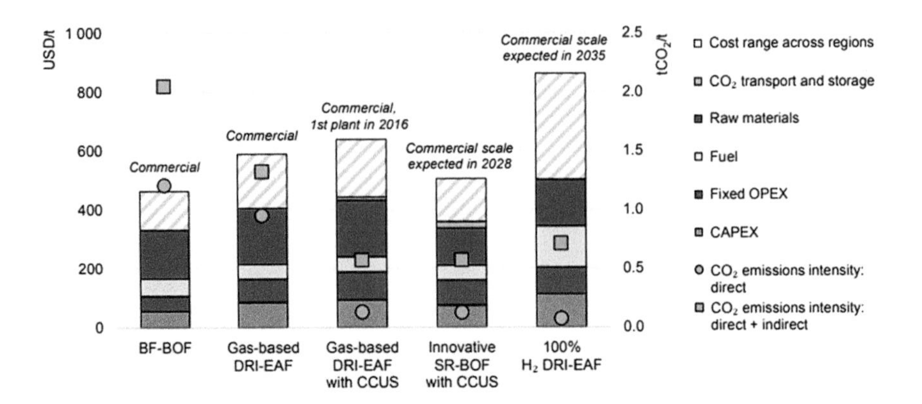

FIGURE 7.8 Estimation of levelized cost of steel production for selected production route. (Source: IEA, Iron and Steel Technology Roadmap Towards more sustainable steelmaking," Part of the Energy Technology Perspectives series, 2020, All rights reserved.)[36] (With permission from IEA, https://www.iea.org/reports/iron-and-steel-technology-roadmap)

of green energy required for these measures, the issue of raising investment and development costs will also be a heavy burden on steelmakers. According to the results of Figure 7.8 based on the survey conducted by IEA, the cost of steel products will almost be doubled if 100% hydrogen reduction and electric furnaces are used.[36] Adoption of CCUS can reduce the cost to a significant level, but this is not an easy task either because the Technology Readiness Level (TRL) of CCUS that handles large amounts of CO_2 is often immature and there are limits to CCUS adoption depending on regional conditions. Therefore, reaching 2050 carbon net zero is expected to be a daunting challenge for major steelmakers. Innovative steelmaking technologies such as hydrogen reduction will be reviewed once more in Chapter 8.

Since more than 2 billion tons of iron ore are mined for the steel industry annually, the impact on the global environment is also very great. Air pollution caused by CO_2, NO_x, PM, etc. emitted from equipment and means of transportation used for mining and transportation should also be considered. However, regarding water pollution that can affect not only mines but also nearby residential areas due to mining, it is necessary to carefully consider the physical and chemical properties of minerals and their impact on water pollution before full-scale mining.

On November 5, 2015, the Fundão Dam in Brazil collapsed, impounding iron ore beneficiation tailings. The accident released more than 40 million m^3 of tailings, killing 19 people and contaminating an area 668 km^2 downstream. The second incident occurred on January 25, 2019, when a tailings dam failed at Córrego do Feijão mine in Brazil, releasing approximately 12 million m^3, which directly affected the administrative area of the company and parts of the communities nearby, resulting in 244 deaths and 26 missing people.[37, 38] As already described in Chapter 3, these accidents primarily affect mining companies that do not have continuous and robust storage methods in terms of civil engineering or environment for tailing that occurs on a large scale for increased steel production. Although the mining company may be responsible, it seems that the steelmaker requesting high-grade ore for productivity cannot be immune from responsibility.[39]

Another type of accident caused by large-scale mining of iron ore is acid mine drainage (AMD). AMD is a phenomenon in which acidic iron compounds such as FeS buried underground react with water during the mining process and change into H_2SO_4, thereby acidifying nearby water. Concerns were raised when water samples taken from a seepage at the Richmond Mine in the United States during restoration efforts in 1990 showed negative pH values, making the water the most acidic ever sampled. Prior to clean-up operations by the Environmental Protection Agency, acid mine drainage from Iron Mountain was among the most acidic and metal-laden anywhere on Earth.[40, 41]

While steel has greatly contributed to mankind's affluent civilized life, it has been used as a war weapon material and had side effects that harmed humanity. Since the Industrial Revolution, many accidents affecting lives and the ecosystem have occurred. After making a huge sacrifice and scientific investigation, it is stepping on the flow of stricter regulations. Once contaminated, it takes a huge amount of money and time to clean up the area. It has become an era where common sense about the pre-regulation environment and the right approach to accident prevention are essential for everyone. All of these were created by humans. In the future, it appears that it is time for us humans, especially those involved in steel, to focus more on research and development so that steel can be used as a means to improve a more comfortable environment and deter war.

Large integrated steelworks generate commensurate amounts of by-products, waste, and wastewater in addition to steel products. The steel industry uses large quantities of water. Nevertheless, a very small amount is consumed since most water is reused or returned to the source. The average water intake for an integrated steelworks is 28.6 m^3/ton of produced steel, with an average water discharge of 25.3 m^3/ton of steel.[42]

Regarding the small amount of wastewater discharged, in most integrated steelworks, it is discharged after undergoing water quality purification according to strict standards.

Valuable metals are recovered as much as possible, and harmful substances are discharged after going through a purification process, depending on the physicochemical characteristics of slag and dust. Blast furnace slag is mostly used as a raw material for fertilizer. Since converter slag has a Fe content of more than 20%, most steel mills are processing it in a way that can recover it as much as possible. The Fe-recovered converter slag is being used as a road pelvis material or fertilizer, but more applications with higher added value should be discovered. Electric arc furnace dust is very high in Zn. Although a technology for recovering this Zn using a carbothermic reaction is commercialized, it is a process that emits CO_2. Recently, technologies that can significantly reduce this CO_2 emission have been reported.[43, 44] However, there are still areas to be improved from the environmental and circular economy viewpoints regarding various types of by-products and wastes generated from large-scale integral steelworks.

REFERENCES

1. John Steele Gordon, "Andrew Carnegie and the creation of U.S. Steel", https://billofrightsinstitute.org/essays/andrew-carnegie-and-the-creation-of-us-steel
2. "The entire history of steel", http://www.matil.org/en/news/steel-sector/the-entire-history-of-steel-1202.html
3. "Charles M. Schwab", last edited March 20, 2024, https://en.wikipedia.org/wiki/Charles_M._Schwab
4. M. S. Birkett, "The iron and steel industry since the war", *J. R. Stat. Soc.*, Vol. 93, No. 3, 1930, pp. 343–397, http://www.jstor.com/stable/2342068
5. Sarah Gilbert, "Bethlehem, Pennsylvania: The town that built America – In pictures", *Wed*, January 18, 2017, https://www.theguardian.com/us-news/gallery/2017/jan/18/bethlehem-pennsylvania-the-town-that-built-america-in-pictures
6. "Empire state building", https://www.pbs.org/wgbh/buildingbig/wonder/structure/empire_state.html
7. Vaibhav Miglani, "The steel industry in USA", 2019, doi:10.13140/RG.2.2.15059.22561, https://www.researchgate.net/publication/334162889_The_Steel_Industry_in_USA
8. Seiichiro Yonekura, "Oshima Takato and the beginning of modem ironmaking", *The Japanese Iron and Steel Industry*, 1850–1990, pp. 18–31, in Japanese, https://link.springer.com/chapter/10.1057/9780230374843_2

9. Kenichi Iida, "Origin and development of iron and steel technology in Japan", *Japanese Experience of the UNU Human and Social Development Programme Series*, Vol. 8, 1980, https://d-arch.ide.go.jp/je_archive/english/society/wp_je_unu8.html
10. A. Fujita, I. Otokozawa, K. Ouken, T. Moriwaki, and Mitoshiro Research Institute, "The light and shadow of the Japanese steel industry", Series, From the Manufacturing Site of Companies and Management 3, 2012, in Japanese
11. Nobuo Ohashi, "Innovation and technical development in the Japanese steel industry", Kawatetsu Techno-Research, Center on Japanese Economy and Business and the Sloan Technology Seminar Series, October 15, 1991, https://www.google.co.kr/url?sa=t&rct=j&q=&esrc=s&source=web&cd=&ved=2ahUKEwj6s6bbjOT8AhXLCYgKHVSaAzU4ChAWegQIDBAB&url=https%3A%2F%2Facademiccommons.columbia.edu%2Fdoi%2F10.7916%2FD8SQ96VH%2Fdownload&usg=AOvVaw2mnpNDIh9Bo7__1Hp5oZFd
12. "The steel industry of Japan", Published by the Japan Iron & Steel Federation, Mar. 1961, https://www.jstage.jst.go.jp/article/isijintoverseas/1/1/1_3/_pdf/-char/en
13. "POSCO's 50 years of history 1968–2018", https://www.posco.co.kr/homepage/docs/eng6/jsp/dn/company/posco/Photo_History_of_POSCO.pdf
14. D. H. Baek, "History of modern Korean steel industry growth", *Iron and Steel Newspaper*, 2010 (in Korean), http://www.snmnews.com/book/bookView.html?idxno=25
15. "History", https://www.hyundai-steel.com/en/aboutus/corporateoverview/history.hds
16. "The most competitive steelmaker in the world 12 consecutive years", World Steel Dynamics (WSD), 2010–2022, https://www.posco.co.kr/brochure/en/01_Overview_03.html
17. "Factbox: A history of China's steel sector", https://www.reuters.com/article/us-china-steel-overcapacity-factbox-idUSKCN0XA03A
18. C. D. Kim, "Steel industry policies under the Xi Jinping administration and their implications", POSCO Research Institute, Asian Steel Watch, Vol. 06, December 2018, https://posri.re.kr/files/file_pdf/82/15577/82_15577_file_pdf_1546391381.pdf
19. Donald B. Wagner, "The earliest use of iron in China", Published in *Metals in Antiquity*, edited by Suzanne M. M. Young, A. Mark Pollard, Paul Budd, and Robert A. Ixer (BAR international series, 792), Oxford: Archaeopress, 1999, pp. 1–9, http://donwagner.dk/EARFE/EARFE.html
20. "Iron making history in China", http://en.chinaculture.org/library/2008-02/01/content_26524.htm
21. Wei Qian and Xing Huang, "Invention of cast iron smelting in early China: Archaeological survey and numerical simulation", *Adv. Archaeomater.*, Vol. 2, No. 1, 2021, pp. 4–14, ISSN 2667-1360, https://doi.org/10.1016/j.aia.2021.04.001.

22. Donald B. Wagner, "Cast Iron in China and Europe", Symposium on Cast Iron in Ancient China, Beijing, July 20, 2009, http://donwagner.dk/cice/cice.html
23. Donald B. Wagner, "Technology as seen through the case of ferrous metallurgy in Han China", http://donwagner.dk/EncIt/EncIt.html
24. "Arcelor Chinese steel venture to start production", March 2005, https://www.bloomberg.com/press-releases/2004-08-27/arcelor-chinese-steel-venture-to-start-production-march-2005
25. "Joint venture between Tata Steel limited and Nippon Steel Corporation", *Press Releases*, Mumbai, January 28, 2010
26. "Japan's JFE to decide on joint venture with JSW in India by 2022-end", September 8, 2022, https://www.business-standard.com/article/companies/japan-s-jfe-to-decide-on-joint-venture-with-jsw-in-india-by-2022-end-122090800535_1.html
27. "POSCO, Indian firm forge partnership", *Korea Times*, August 16, 2007, https://www.koreatimes.co.kr/www/tech/2022/10/129_8485.html?utm_source=KK
28. "Arcelor Mittal and Nippon Steel complete acquisition of Essar Ssteel", December 16, 2019, https://corporate.arcelormittal.com/media/press-releases/arcelormittal-and-nippon-steel-complete-acquisition
29. "We are a joint venture between Indonesia and Korea", https://www.krakatauposco.co.id/about
30. "The most competitive steelmaker in the Southeast Asia No. 1", https://www.fhs.com.vn/Portal/home
31. "Tata Steel Europe", last edited February 7, 2024, https://en.wikipedia.org/wiki/Tata_Steel_Europe
32. "ThyssenKrupp", last edited March 24, 2024, https://en.wikipedia.org/wiki/ThyssenKrupp
33. "Evolution of ArcelorMittal in the world", https://rails.arcelormittal.com/about-us/history
34. "AM/NS Calvert", https://northamerica.arcelormittal.com/our-operations/am-ns-calvert
35. "ArcelorMittal completes transaction to acquire Ilva S.p.A. and launches ArcelorMittal Italia", January 2018, https://corporate.arcelormittal.com/media/press-releases/arcelormittal-completes-transaction-to-acquire-ilva
36. "Iron and steel technology roadmap towards more sustainable steelmaking", Part of the Energy Technology Perspectives series, International Energy Agency (IEA), 2020, https://aceroplatea.es/docs/Iron_and_Steel_Technology_Roadmap_IEA.pdf
37. C. dos SantosVergilio, D. Lacerda, B. deOliveira, E. Sartori, G. M. Campos, A. L. de S. Pereira, D. B. deAguiar, T. da S. Souza, M. deAlmeida, F. Thompson, and C. E. de Rezende, "Metal concentrations and biological effects from one of the largest mining disasters in the world (Brumadinho, Minas Gerais, Brazil)", *Nat. Res. Sci. Rep.*, No. 10, 2020, p. 5936, https://doi.org/10.1038/s41598-020-62700-w

38. E. T. Parente, Adan S. Lino, Gabriel O. Carvalho, Ana C. Pizzochero, Claudio E. Azevedo-Silva, Matheus O. Freitas, Cláudia Teixeira, Rodrigo L. Moura, Virgílio José M. Ferreira Filho, and Olaf Malm, "First year after the Brumadinho tailings' dam collapse: Spatial and seasonal variation of trace elements in sediments, fishes and macrophytes from the Paraopeba River, Brazil", *Environ. Res.*, Vol. 193, February 2021, p. 110526, https://doi.org/10.1016/j.envres.2020.110526
39. S. W. Branch, "Extraction and Beneficiation of Ores and Minerals Volume 3 Iron", *Technical Resource Document*, August 1994, U.S. Environmental Protection Agency, https://www.google.com/url?sa=t&rct=j&q=&esrc=s&source=web&cd=&cad=rja&uact=8&ved=2ahUKEwjzv9f2gKeBAxWnfPUHHV-vBewQFnoECBgQAQ&url=https%3A%2F%2Farchive.epa.gov%2Fepawaste%2Fnonhaz%2Findustrial%2Fspecial%2Fweb%2Fpdf%2Firon.pdf&usg=AOvVaw0B_71qsvI9JgSYkw4OQSfd&opi=89978449
40. F. W. Mammano, "Iron mountain: The history and complications in metal ore mining, remediation and reclamation of a national superfund site", CSUS ENVS 190 –Thesis, 2018, https://www.csus.edu/college/social-sciences-interdisciplinary-studies/environmental-studies/_internal/_documents/iron-mountain-the-history-and-complications-in-metal-ore-mining,-remediation-and-reclamation-of-a-national-superfund-site---frank-william-mammano.pdf
41. Ata Akcil and Soner Koldas, "Acid Mine Drainage (AMD): Causes, treatment and case studies", *J. Clean. Prod.*, Vol. 14, 2006, pp. 1139–1145
42. V. Colla, I. Matino, T. A. Branca, B. Fornai, L. Romaniello, and F. Rosito, "Efficient use of water resources in the steel industry", *Water*, Vol. 9, 2017, p. 874, doi:10.3390/w9110874
43. P. Palimaka, S. Pietrzyk, M. Stepien, K. Ciecko, and I. Nejman, "Zinc recovery from steelmaking dust by hydrometallurgical methods", *Metals*, Vol. 8, 2018, p. 547, doi:10.3390/met8070547
44. R. Chairaksa, Y. Inoue, N. Umeda, S. Itoh, and T. Nagasaka, "New pyro metallurgical process of EAF dust treatment with CaO addition", *Int. J. Miner. Metall. Mater.*, Vol 22, January 8, 2015, pp. 788–797, doi:10.1007/s12613-015-1135-6

CHAPTER 8

Steel Industry, Future

PREDICTING THE FUTURE IS never easy. However, it is important to attempt to envision what may happen in the future and anticipate the potential challenges we might face. We need to assess the seriousness of these issues and explore practical solutions. In previous chapters, we have explored the significant role that iron and steel have played in shaping human life, civilization, culture, and thought over the past 5,000 years. Therefore, it is essential and highly intriguing to examine how the steel industry itself will evolve in the future. In this chapter, we will delve into four key areas: the steel manufacturing process, products, the industry, and smart factories to understand the changes that lie ahead.

8.1 THE FUTURE OF STEEL MANUFACTURING PROCESSES

Iron has been utilized by humanity for over 4,000 years, and it will remain an integral part of our society for as long as humankind exists. Figure 8.1 illustrates the global trend in steel production.[1] Figure 8.1 shows an average annual growth of 2.9% over the past 60 years. Notably, China and India have experienced rapid increases in steel consumption over the past two decades and 10 years, respectively. Moreover, countries such as Indonesia, Pakistan, Nigeria, Bangladesh, Mexico, Vietnam, and Ethiopia, with their large populations, are expected to witness a significant surge in steel consumption soon.

One notable transformation in the steel industry during the past 60 years has been the integration of processes to facilitate mass production and reduce manufacturing costs. As discussed in Chapter 7, the success story of Japan, Korea, and China can be attributed to the implementation

DOI: 10.1201/9781003419259-8

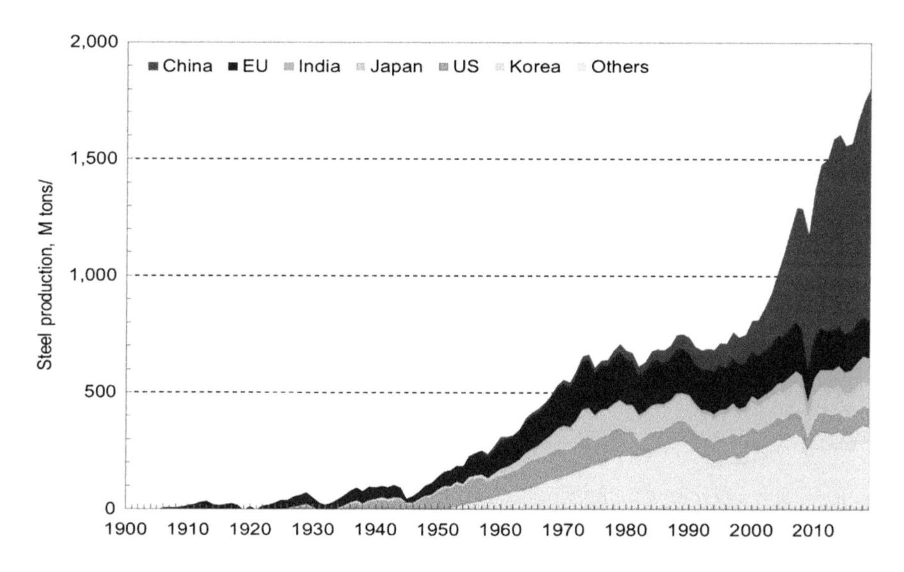

FIGURE 8.1 Steel production trends in major steel-producing countries since 1900.[1]

of integrated steelworks systems, where all processes from raw material unloading to product shipment are carried out within a single region. However, even within this integrated system, as depicted in Figure 1.5, major processes such as coking, sintering, and BF operations have been subject to short-circuiting within the entire plant. Therefore, efforts to further integrate these processes have been continuously pursued.

8.1.1 New Ironmaking Process

Currently, the ironmaking process involves separate coking and sintering processes preceding the BF. The BF is a typical emitter of CO_2,[2] and both coking and sintering generate substantial amounts of CO_2, SO_x, NO_x, and fine dust particles.[3, 4] Over the past 40 years, various alternatives, collectively known as the new ironmaking process, have been developed to address these issues and integrate these processes. These new ironmaking processes can be categorized based on the reducing agent and the particle size of the ore, and the currently developed cases are presented in Table 8.1.[5–11]

Although some of these new processes are currently in commercial operation, they have not completely replaced the BF due to limitations in productivity, yield, and quality. As a result, these new processes are currently implemented on a smaller scale in regions where favorable raw materials and local conditions exist. The MIDREX process is considered

TABLE 8.1 New Ironmaking Process Under Development or Commercialization

Reductant	Size of ore	Process	Company	No. of facilities/ Capacity, MT/y	Commercial
Coal	Pellet/ Lump	COREX	Primetals	JSW, 2/ 0.6 ESSAR, 1/0.6 Saidanha, 1/0.6 Baowu, 1/1.5	Yes
	Fine	Hismelt	Rio Tinto	Molong, 1/0.6	Demo
		Hisarna	Tata EU		Pilot
		FINEX	POSCO/ Primetals	1/1.5 1/2.0	Yes
		ITmk3	Kobe	US 1/0.5	Yes
		FASTMET	Kobe	6/0.2	Yes
Gas	Fine	FINMET	Primetals	4/0.5	Stop
	Pellet/ Lump	HYL (ENERGIRON)	Teronva/ Danieli/ NSENGI	32/27.0	Yes
		MIDREX	Kobe-Danieli	73/58.3	Yes

advantageous in areas with abundant natural gas supply,[5] while the FINEX process is suitable for inland regions with large quantities of low-grade iron ore.[9] However, the prospects for these new processes are promising, as the reserves of hard coal for coking are depleting rapidly, and there is an increasing focus on regulations concerning CO_2 and fine dust emissions. With stricter carbon regulations, it will be challenging to address CO_2 emissions in the BF process unless CO_2 can be separated and stored at a reasonably low cost. However, since the current reducing agent used in the new ironmaking process is more expensive than coal used in the traditional BF process, innovative cost-effective processes need to be developed to ensure their competitiveness.

Since the 1980s, COREX plants with an annual capacity of 300,000 tons have begun operations in South Africa and POSCO. However, COREX had a limitation that required the use of pelletized or sintered iron ores and coals. In 1998, POSCO and VAI (now Primetals) started developing FINEX, a new ironmaking process that utilizes 100% fine-size coal and ore.

Figure 8.2 provides an overview of the FINEX process. The fine ore undergoes a reduction in the fluidized bed reactor (FBR) using the gas

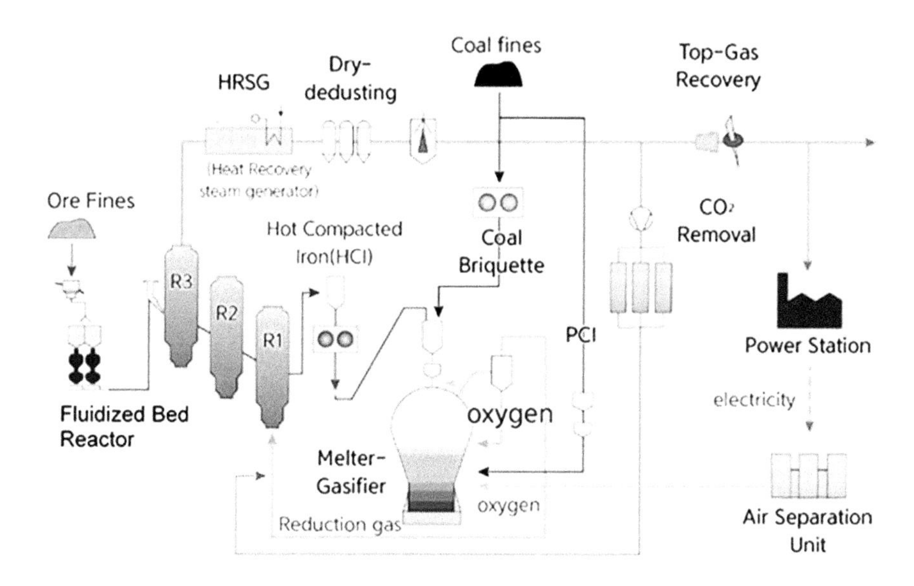

FIGURE 8.2 Outline of POSCO FINEX process.

released from the melter gasifier (MG) to produce hot compacted iron (HCI). HCI is then introduced into the MG along with a coal briquette, resulting in the formation of molten pig iron. In the MG, coal briquettes and pulverized coal are burned to generate a reducing gas, which is supplied to the FBR to carry out the reduction of fine ores in three stages. Some of the gas discharged from the FBR undergoes treatment in the CO_2 removal facility to reduce the CO_2 content to 3% or lower. This treated gas is mixed with the MG gas to aid in the reduction of HCI in the FBR. The remaining off-gas is utilized as fuel for top gas recovery and off-gas power generation. The FINEX process produces reduced amounts of off-gas and dust compared to the BF process due to its integrated closed-circuit design and utilization of pure oxygen. The quantities of NO_x, SO_x, and dust generated in the flue gas are 40%, 15%, and 70% of those produced by the BF process, respectively.

The key technology of POSCO FINEX lies in the production of coal briquettes capable of high-speed feeding while maintaining their shape even under high-temperature conditions. POSCO initiated the production of 1.5 million tons/year of FINEX in 2007 as the first phase, and in 2014, they added a second phase FINEX facility with a capacity of 2 million tons/year. It was discovered that the manufacturing cost was comparable to that of the BF process at the same production volume. The FINEX process enables CO_2 capture and storage (CCS) through the CO_2 removal facility, resulting in a reduction of CO_2 emissions by over 35%. As environmental regulations become more stringent, the value and competitiveness of the FINEX process are expected to further increase.(12)

The MIDREX process is the world's most widely used commercial new ironmaking method. This process, which was developed in the United States in the 1960s, produces direct reduced iron (DRI) by gas–solid reaction by injecting natural gas and oxygen into the lower part of the shaft furnace, as shown in Figure 8.3.(13, 14) This process is expanding its application in the United States, Mexico, India, and Iran, where natural gas can be supplied inexpensively. Looking at the reaction mechanism, it is the authors' opinion that this process can be seen as an extension of the bloomery furnace we used in ancient times. Gas is supplied instead of charcoal, and the reduction reaction of iron ore by the natural gas is the same

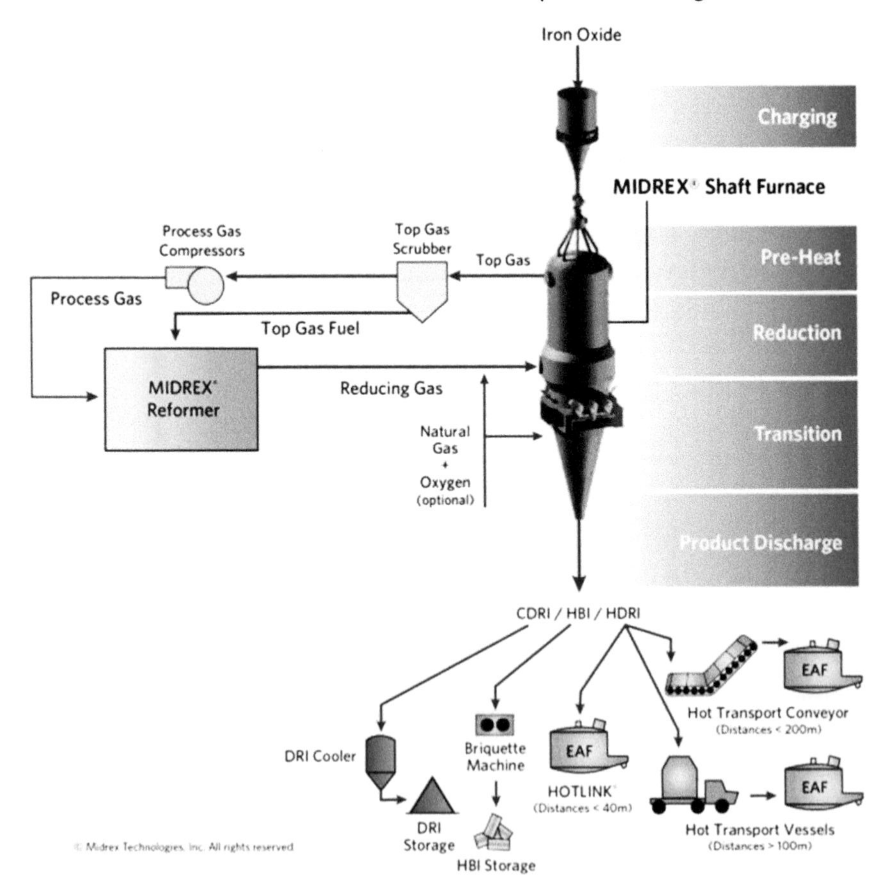

FIGURE 8.3 Outline of MIDREX process. (With permission from MIDREX, https://www.midrex.com/technology/midrex-process/)

in principle. Natural gas, of which CH_4 is the main component, involves some hydrogen reduction reaction. Since MIDREX produces solid DRI in a shaft furnace, an additional electric furnace is required to melt this DRI. The MIDREX process has undergone many improvements over the decades. Recently, the capacity of the shaft furnace has been increased to produce up to 2 million tons/year, and the high-temperature DRI from the shaft furnace is directly charged into the electric furnace by hot transport. In addition, a method of reducing ore using hydrogen alone is also being developed. A steelmaker operating the MIDREX process has the advantage of being able to change the reducing agent from natural gas to hydrogen without significantly modifying the facility. Therefore, ArcelorMittal, Nippon Steel, Baowu, USS, SSAB, etc. are either planning or constructing the shaft furnace–type hydrogen reduction facilities.

8.1.2 Thin Slab Casting and Strip Casting

Attempts to integrate the CC and hot-rolling processes have been ongoing for the past 50 years, with two notable examples being the TSC process and the SC process (see Figure 8.4). In the TSC process, casting and rolling are not fully integrated, as the cast strands from two casting machines are rolled in a single facility through a tunnel furnace. However, since the 2000s, an integrated process has been developed that eliminates the need for tunnel furnace heating. POSCO has successfully commercialized the

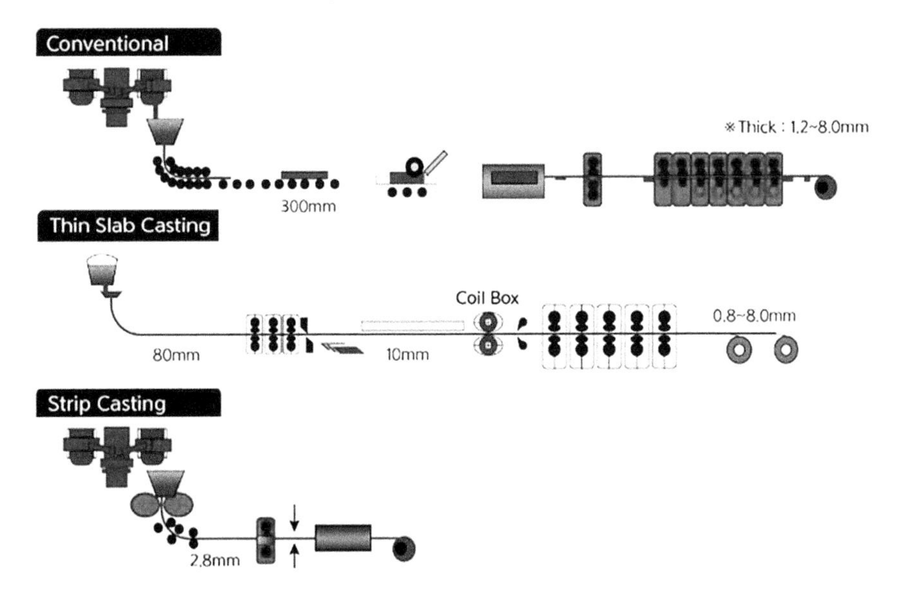

FIGURE 8.4 Comparison of the process integration of CC and hot rolling.

Compact Endless Casting and Rolling Mill (CEM) process as an integrated casting-hot rolling process, while Arvedi has developed the Endless Strip Production (ESP)[15] process as a similar technology. Both CEM and ESP achieve seamless process integration by continuously feeding the casting line to the finishing mill without any short-circuiting. This approach offers the advantage of eliminating quality deterioration issues in the top and tail sections during hot rolling, as well as significantly reducing capital investment costs. Achieving high-speed casting is a prerequisite for matching the rolling speed. It has been reported that the CEM line at POSCO has achieved the world's highest casting speed of 8.0 m/min.[16]

The CEM process at POSCO was successful in the demonstration stage but faced limitations, such as product width, which halted its further expansion. On the other hand, ESP has been adopted by Chinese steelmakers like Rizhao Steel (日照鋼鐵), specializing in thin-gauge products that are challenging for existing hot rolling processes to manufacture.[17] The existing TSC process has a capacity of 2 million tons/year, but CEM allows for a 4 million tons/year plant. USS invested US$3 billion in constructing two EAFs and two ESPs at Big River Steel Works in Osceola with a production capacity of 6.3 million tons/year, Arkansas, to enhance its cost competitiveness in the US.[18] Both CEM and ESP processes are prospective options for future greenfield or rationalization investments due to their significant reduction in CO_2 emissions by eliminating BFs.

Since the 1980s, NSC, TKS, POSCO, Baowu, and Nucor have made attempts to commercialize the SC process.[19] Nucor successfully commercialized the Castrip process, a twin-roll casting technology developed by Australia's BHP Steel and Japan's IHI, at two facilities in Indiana and Arkansas.[20] It has been reported that Shasteel, an integrated steelworks, has also installed Castrip and is operating it successfully.[21] The extent to which the company has achieved cost and quality goals is unknown. Other companies, including POSCO, have abandoned the development of the SC process for stainless steel. While the quality level of stainless steel produced through strip casting is comparable to conventional steel,[22] it has not yet achieved price competitiveness with conventional processes primarily due to the limited lifespan of expensive edge dams.

8.2 THE FUTURE OF STEEL PRODUCTS

New steel products are constantly being developed, and innovative products based on new metallurgical theories are emerging. The development of these innovative steel products is driven by two main factors. The first is customer demand, and the second is the development of new theories. In the first case, when a customer requires a new product with innovative characteristics during the design process, the steel company undertakes the development and supply of the new product to meet the customer's needs. The second case involves the creation of products with new properties based on new theories proposed by universities or research institutes regarding materials. Examples of such products include dual-phase (DP) steel, transformation induced plasticity (TRIP) steel used in automobiles, and the recently developed twinning-induced plasticity (TWIP) steel. These products fall into the category of being developed based on new material theories.

The development of new products begins with laboratory experiments that stimulate the manufacturing conditions and evaluate the performance of the test coupons. Michael Faraday (1767–1867) may be the first who built laboratory equipment and made test specimens. Using his blast furnace, he manufactured test coupons to evaluate the effects of alloying on their properties, pioneering metallurgical research.(23) The optical microscope was a widely used tool for microstructural analysis and began to be applied for specimen analysis by Henry Clifton Sorby (1826~1908). In 1886, Sorby published a paper on the lamellae structure of pearlite, and suggested the influence of alloying on the structure and hardenability of martensite, providing insight into the hardenability of alloy steels.(24) Robert Abbott Hadfield (1858~1940) who became managing director of a casting company had invented a high Mn wear-resistant steel which is still used widely these days. Hadfield's interest in alloy steels was not confined to Mn-containing steel. He was active in developing various special steels for armor-protecting and heat-resisting purposes. As a result, he was the recipient of several prestigious awards including the Bessemer Gold Medal.(25) Another example of metallurgical research is the invention of stainless steel by Harry Brearley (1871~1948), who created steel with 12.8% Cr and 0.24% carbon. All of these R&D efforts for commercialization have laid the basis for new future steel development with wide applications.(26)

8.2.1 Future Steel for Car Body

There is a global trend towards strict regulation of vehicle exhaust emissions.(27) As a result, the automotive industry is not only focusing on improving power transmission and transitioning to eco-friendly vehicles such as electric cars but also competing in the design of lightweight car bodies through the use of high-strength steel materials. Various methods, including solid solution hardening explained in Chapter 1, are being considered to strengthen steel for car bodies.

In the past, interstitial free (IF) steel with a TS of 300 MPa or lower was primarily used for the outer parts of automobiles. However, in recent years, DP and TRIP steels with TS values of 490 or 590 MPa have been applied by incorporating principles such as bake hardening, solid solution hardening, precipitation hardening, and transformation hardening.[28] Looking ahead, it is expected that the TS values will increase to approximately 780 or 980 MPa.

For the inner parts of car bodies solid-solution and precipitation-hardened high-strength steels have been used as lightweight materials, and the use of transformation-hardened high-strength steels is expanding. Recently developed steel materials have achieved TS values exceeding 1.0 GPa, and many "Giga steel" products have been developed.[29] Compared to other lightweight materials like aluminum, Giga steel is receiving significant attention as a futuristic lightweight material. To achieve weight reduction in vehicles, good formability is essential. Figure 8.5, known as the banana diagram, illustrates the relationship between TS and elongation (El) for various steel materials. If a steel material is positioned in the top-right section of the banana diagram, it is considered to have high strength and excellent workability.

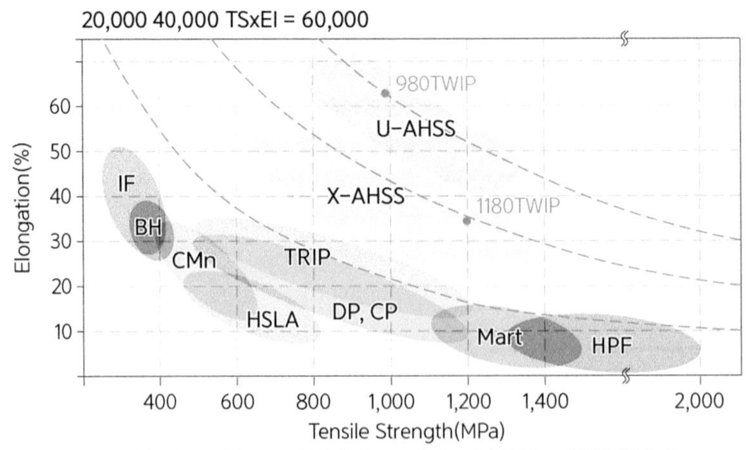

FIGURE 8.5 High-strength steels for car bodies are indicated on the banana diagram.

U-AHSS and X-AHSS, located in the upper-middle section of Figure 8.5, correspond to Giga Steel, and commercialized variants such as 980TWIP and 1180TWIP have been introduced.[30] The TSxEl (tensile strength times elongation) value of these steel falls within the range of 40,000 to 60,000 MPa·%, and they exhibit superior formability compared to other high-strength steels. Hot press forming (HPF) is a combination process involving heat treatment and stamping for the formation of automotive parts and is widely employed for ensuring structural stability. However, a drawback of HPF is the formation of an oxide film on the surface due to high-temperature heat treatment, resulting in an uneven surface. To address this issue, the warm press forming (WPF) method can be utilized by optimizing the alloy composition to lower the heat treatment temperature. Giga steels like U-AHSS and X-AHSS are particularly effective in lightweighting the car body when relatively simple-shaped car parts are fabricated at room temperature.

TWIP steel, initially developed by POSCO for car bodies, is high-manganese steel with a composition of 0.6C–18Mn–1.5Al. It belongs to the Giga steel category and exhibits ultra-high strength and high formability.[28, 30–33] Similar steel grades called X-IP and L-IP have also been developed by AM,[34] TKS,[35] and Baowu.[36] In TWIP steel, the stacking fault energy (SFE) plays a crucial role in determining the microstructure of the material. Plastic deformation of this steel leads to the formation of twins, which enhances the work-hardening effect while maintaining formability. This phenomenon is known as the TWIP phenomenon.[31]

Figure 8.6 compares the TSs of DP and TRIP steels with the same strength as TWIP steel (TS 980MPa), highlighting that TWIP steel exhibits much higher elongation. The manufacturing cost of TWIP steel is higher than that of TRIP steel due to the larger amount of alloying required and the lower yield caused by quality defects. However, the prospects are promising as the defect rate and cost have significantly decreased with recent advancements in process technology.

The term "Giga steel" mentioned earlier holds special significance in terms of lightweight. Since the TS of Giga steel is more than three times that of aluminum, the specific strength obtained by dividing the TS by the specific gravity of Giga steel exceeds that of aluminum. Consequently, if automotive structural parts serving the same purpose are made of Giga steel, their weight can be reduced compared to aluminum counterparts, leading to substantial cost savings. While some automakers tend to prefer aluminum for car bodies due to its lightweight nature, the discovery of

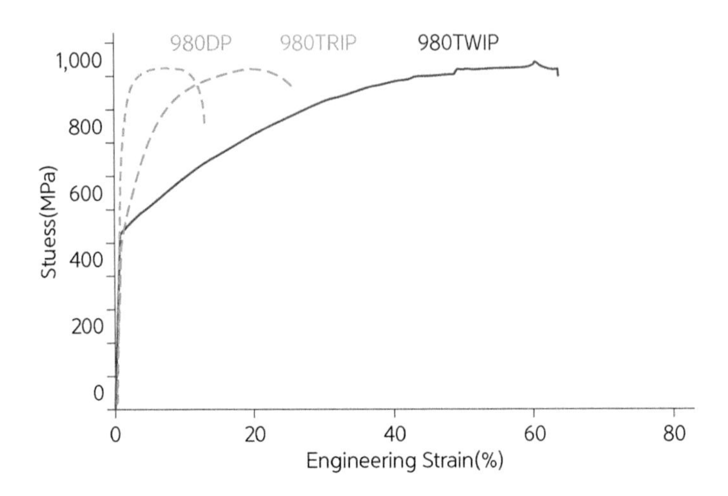

FIGURE 8.6 Tensile curves of DP, TRIP, and TWIP steels with TS of 980 Mpa.

Giga steel will prompt them to reconsider their choices. Utilizing Giga steel allows automobile bodies to benefit from both reduced manufacturing costs and weight, thanks to good weldability and decreased assembly time for parts. As the proportion of electric vehicles increases in the future and automotive structural parts adopt simplified designs, the utilization of Giga steel will become even more prevalent.[32]

8.2.2 Future of High Mn Steel

High Mn steel is emerging as a next-generation steel material. High Mn steels containing 10 to 24% Mn were commercialized by Sir. Hadfield in 1882.[25, 37] Hadfield steel was used in bulletproof helmets and heavy equipment caterpillars during World Wars I and II, and it is still employed for military purposes such as crushers, basket blades, rails, and rail points due to its excellent abrasion resistance. Additionally, 3% Mn steel has been developed for heavy construction equipment that requires wear resistance.[30]

A new application of high Mn steel is seen in the use of 18% Mn steel for slurry pipes in oil sand fields. A major US oil company recently replaced existing line pipe steel with this high Mn steel for oil sand transportation. Despite the higher cost of high Mn steel compared to ordinary line pipe steel, its superior abrasion resistance justified the investment. The economic benefits were significant, as the replacement cost of pipelines in inaccessible locations was substantially reduced. The extended lifespan of the high Mn steel pipes resulted in considerable cost savings, about three times the previous lifespan.[30, 38, 39]

Another future application of high Mn steel lies in the LNG-related field, where demand is expected to grow. As mentioned in Chapter 2, steels tend to become brittle at low temperatures. However, steels with an austenite structure can exhibit excellent toughness even at cryogenic temperatures, which is crucial for handling LNG. The use of LNG as an alternative to coal or oil is increasing due to environmental concerns. In the case of steel for LNG storage tanks, 304 stainless steel and 9% Ni steel with an austenite structure are commonly used. Manganese (Mn) has a strong austenite stabilizing effect similar to nickel (Ni). Since the price of Mn is less than 1/8 of Ni as of 2022, it has garnered significant attention from customers in terms of cost-effectiveness.[40]

When considering normal material costs and processing conditions, such as welding materials suitable for each material, a high-Mn steel tank can reduce manufacturing costs by more than 10% compared to a tank made of 304 stainless steel or 9% Ni steel. The certification of high Mn steel in various international standards related to cryogenic use has led to an increase in purchasing orders from shipbuilders and shipowners.[41]

Figure 8.7 compares the results of the Charpy V-Notch impact test for four materials used in cryogenic applications at different temperatures. The impact value of high Mn steel is similar to that of 9% Ni steel and significantly higher than that of 304 stainless steel. Figure 8.8 illustrates the tensile properties of the four materials. At room temperature, the tensile

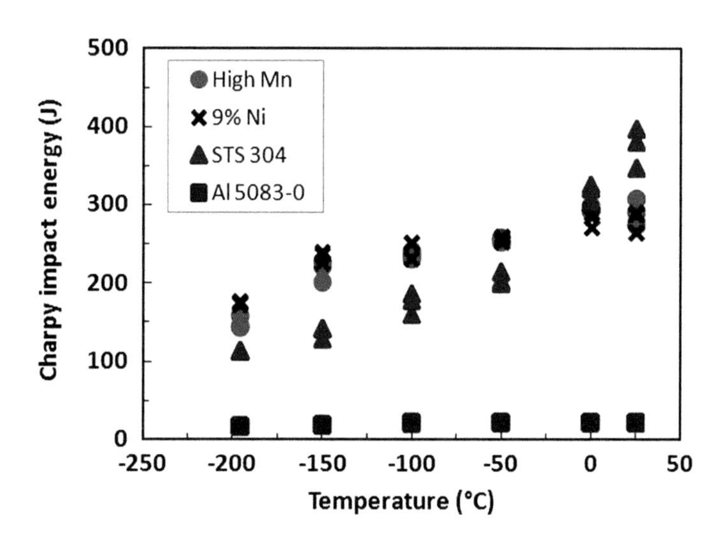

FIGURE 8.7 Charpy impact energy levels of four different cryogenic materials.[42]

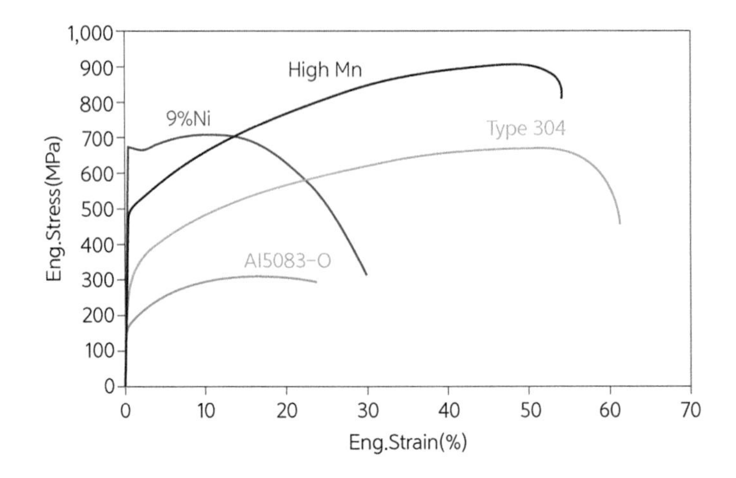

FIGURE 8.8 Tensile properties at 25°C of high Mn steel, 9% Ni steel, 304 Stainless, and Al alloy, which are materials for cryogenic use.[42]

elongation of high Mn steel surpasses that of 9% Ni steel, indicating its superior workability.[42]

Although numerous high-strength steel products are currently under development, it is challenging to find steels with as many possibilities as high-Mn steels, including excellent mechanical properties. High Mn TWIP steel is primarily produced for car bodies, while high Mn steel with a martensitic structure is mainly used for plates. However, it will certainly be applied not only in civil engineering, construction, mechanical structures, pressure vessels, and pipes but also in aerospace applications, where expensive alloys are typically employed. High Mn steel with an austenitic structure exhibits exceptional low-temperature toughness, making it suitable for LNG containers and similar purposes. The commercialization of high-Mn steel for various applications will require securing cost competitiveness by reducing the additional cost burden associated with changes in the manufacturing process, and continuous development plans should be formulated for this purpose in the future.

8.2.3 Future of Functional Steel – Surface Treated Steel

Corrosion is a material disadvantage for iron due to its susceptibility to oxidation and tendency to revert to iron ore components in the presence of oxygen and moisture. However, properly surface-treated iron can overcome this disadvantage and acquire new functionalities. Surface treatment of steel can be categorized into a coating process, which involves applying a metal with a higher ionization tendency than iron onto the surface, and a post-treatment coating, where a function is imparted by treating the surface with a chemical solution or similar methods. Zinc, known for its sacrificial anodic behavior and excellent corrosion resistance, is commonly used as a coating material. Zinc is a non-ferrous metal with abundant resources and a low melting temperature, making it advantageous in terms of requiring less coating energy. For example, if the zinc coating amount on a currently commercially available steel sheet is 60 g/m^2, corrosion can be prevented for approximately 10 years.[43, 44]

Hot dip galvanizing and electro-galvanizing (EG) are the two commercialized methods for zinc coating. Additionally, steel sheets coated with a composite alloy of aluminum and magnesium have been developed to achieve excellent corrosion resistance through the formation of self-oxidized films.[45–47] Figure 8.9 shows that the corrosion resistance index (CRI) tends to increase with the amount of aluminum and magnesium added.[48] However, increasing the magnesium content results in a strong oxidizing property, which poses challenges in the coating process. Nevertheless, recent advancements in anti-oxidation technology in the coating bath have enabled the release of products with up to 5% magnesium content.[49]

To overcome the limitations of alloy addition in hot-dip galvanizing, advanced technologies like physical vapor deposition (PVD) have been explored. Steelmakers have been attempting to commercialize this technology for the past 20 years, and companies like AM and POSCO are operating demo plant-level facilities.[50, 51] According to Figure 8.9, the CRI of PVD steel sheets, which contain up to 20% magnesium in the

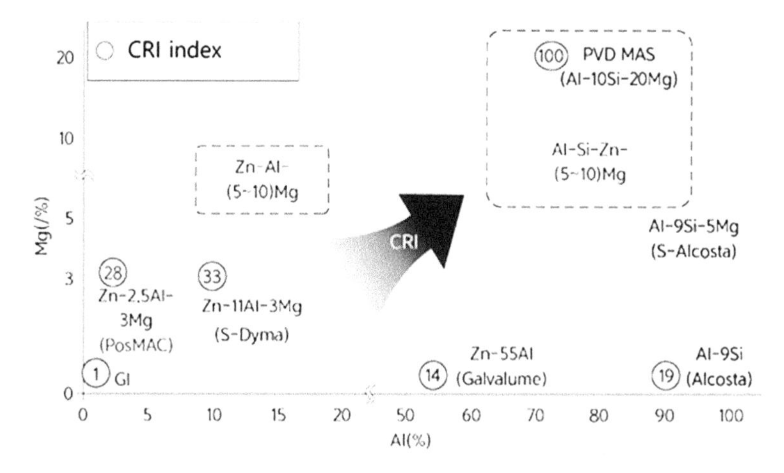

FIGURE 8.9 Recently developed Mg–Al-based alloy coated steel sheets and CRI.[48]

coating layer and added silicon, surpasses that of conventional coated steel sheets.[52] Additionally, vacuum deposition equipment has a lower investment cost compared to hot-dip galvanizing, providing an advantage in terms of equipment investment.

Various functionalities can be achieved through post-treatment of the coated surface. These treatments include phosphate treatment to enhance paint adhesion, chromate coating to improve corrosion resistance, resin coating to enhance fingerprint resistance or lubricity, and organic/inorganic composite coating to improve fuel corrosion resistance and weldability. Currently, there are ongoing efforts to develop organic-inorganic composite coating technologies with diverse functionalities.

Some examples of these technologies include the composite resin hybrid GI steel sheet, which combines organic and inorganic materials,[53] eco-friendly ultra-violet (UV) curing coated steel sheet,[54] high heat dissipation composite resin steel sheet,[55] decorative steel sheet utilizing inkjet printing,[56] antibacterial coating technology,[57] and more. Furthermore, in the future, we can expect the emergence of surface-treated steel sheets such as organic-inorganic composite coated flame-retardant steel sheets for building materials,[58] highly workable nano-patterned steel sheets,[59] and heat-sensitive color-coated steel sheets.[60]

8.3 THE FUTURE OF THE STEEL INDUSTRY

8.3.1 Future Demand for Steel

The rapid proliferation of electric vehicles in the automobile industry, which is one of the major consumers of steel, raises intriguing questions about the potential changes in the steel market. Tony Seba predicted a significant decrease in automobile demand due to the rise of the sharing economy and self-driving cars.[61] However, it is generally believed that the demand for cars will continue to grow in the future. This is because businesses operating car-sharing systems are expected to increase their vehicle ownership, leading to a significant rise in the overall mileage of vehicles worldwide.

On the supply side, the overcapacity in steel production has become a global concern in the steel industry. Figure 8.10 illustrates the historical trend of the overcapacity and the utilization rates over the past 60 years. It reveals a recurring pattern of fluctuations within the range of 80±10%, depending on the international economic environment. In the 2000s, the massive expansion of steel facilities in China led to a critical issue of overcapacity within the OECD.[62]

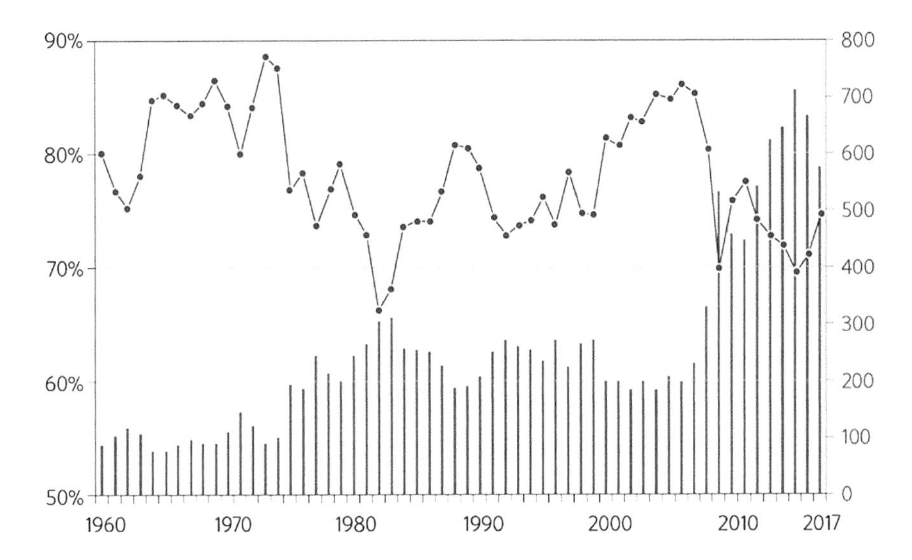

FIGURE 8.10 Overcapacity in the global steel industry. (The bar in the graph is overcapacity in million tonnes on the right y-axis and the line with closed circle is the operation ratio (actual production/capacity) on the left y-axis.)

Recently, China announced the closure of more than 100 million tons of steel facilities. However, these closures primarily affected unquantified Titiaogang companies. Additionally, large BF companies in China have improved their production efficiency through mergers and technological advancements, leading to an overall increase in production. Consequently, the global overcapacity issue persists, with a total excess capacity of more than 600 million tons. If we consider this problem on a national level, particularly since steel is regarded as a vital industry, the issue of overcapacity will persist until struggling companies are phased out.

Red Beddows predicts that global steel demand will surpass 4 billion tons by 2050, more than double the production levels of 2018.[(63)] Based on the steel demand of 1.81 billion tons in 2018, this suggests an average annual growth rate of 2.5%. Beddows assumes that the growth rate of steel demand will be around 70% of the United Nations' world GDP growth rate. However, an average annual growth rate of 2.5% may seem overly optimistic considering factors such as emerging CO_2 emission regulations and population decline.

The rapid growth in steel demand is primarily driven by national infrastructure construction and the expansion of manufacturing, particularly in countries with a per capita GDP ranging from US$5,000 to US$20,000. However, as many countries that have already reached a developed stage experience a shift from manufacturing to the service industry in their industrial structure, it becomes challenging to sustain the 2.9% growth rate observed over the past 60 years as shown in Figure 8.1. Future GDP growth, especially, will heavily rely on the service industry, sharing economy, and the Fourth Industrial Revolution, rather than manufacturing. Consequently, the growth rate of steel demand is likely to decelerate.

According to data reported by Accenture to the OECD in 2017,[(64)] taking into account negative factors like the sharing economy and positive factors such as explosive growth in India, steel demand in 2035 was projected to reach 1.87 billion tons with a growth rate of 1.1%. However, recent years have seen steel demand surpass these projections, with growth rates of 3.6% in 2018, reaching 1.8 billion tons, and 1.9 billion tons in 2021.

Population growth is a crucial variable among the various factors influencing steel demand. In the case of developing countries like India, it is widely believed that global steel consumption will continue to rise for several decades due to the relatively high proportion of young people and the need for infrastructure development. Taking these factors into account, it is predicted that the growth rate of steel demand will fall between Accenture's 1.1% and Beddows' forecast of 2.5%. Assuming a growth rate of 1.8%, which lies halfway between these two projections, steel demand in 2050 is estimated to reach 3.2 billion tons.

For the steel industry to achieve sustainable growth, continuous research and development (R&D) efforts are necessary to develop new products with enhanced functionalities, while maintaining a cost advantage over alternative materials and minimizing adverse environmental impacts.

8.3.2 Environment-friendly Steel Industry – Hydrogen Reduction and Others

One of the readily available methods to reduce carbon dioxide emissions is the utilization of steel scrap as a source of iron. Currently, approximately 650 million tons of steel scrap are used annually. The volume of end-of-life scrap is expected to increase significantly, with projections of 600 million tons by 2030 and 900 million tons by 2050.[(65)] It has been noted that steel products manufactured from scrap may have lower quality compared to those produced from BF materials due to the presence of tramp elements. However, for automotive steels and other products that tolerate a minimal amount of impurities, EAFs can be employed, as demonstrated by Nucor's example.[(66)] Strategies such as incorporating tramp elements as alloying elements or removing them during the steelmaking process should be developed.

As long as the BF is in operation, the reduction of iron ore using carbon leads to the inevitable emission of greenhouse gases. Therefore, the utilization of hydrogen as a reducing agent is being prioritized. Although the use of natural gas as a reducing agent, combined with coal, in new ironmaking processes can partially reduce carbon dioxide emissions, the fundamental solution lies in the adoption of hydrogen as the primary reducing gas. Consequently, major steel mills worldwide, such as SSAB,(67) AM,(68) POSCO,(69) Nippon Steel,(70) and Baowu,(71) are actively engaged in the development of hydrogen reduction (HR) steel technology.

There are several challenges associated with the development of hydrogen reduction technology. First, the HR process for iron ore is an endothermic reaction, meaning that additional energy must be supplied externally.(72) Second, a substantial amount of electric energy is required to produce the hydrogen used for reduction.(73) Third, hydrogen-reduced iron exists as a solid phase containing partially reduced iron oxide, necessitating the use of an electric furnace capable of melting it. Fourth, since the solid iron produced through hydrogen reduction contains very little carbon, an additional heat source is required for the subsequent steelmaking process. Consequently, in all of the aforementioned cases, the energy supplied must come from electricity generated using renewable green energy sources.

As mentioned earlier, the hydrogen reduction process requires a significant amount of energy compared to the conventional coal-reduced BF, and this energy must be generated from renewable sources. Typically, about 50 kWh of electricity is required to produce 1 kg of hydrogen,(73) and 54 kg of hydrogen is needed to reduce Fe_2O_3 to 1 ton of Fe. Based on the unit price of power generation in Korea in 2021, nuclear power costs approximately US$2.2 to produce 1 kg of hydrogen.(74) Coal is 1.8 times more expensive than nuclear power, LNG is 2.2 times more, and solar/wind power is 1.9 times more. If all 47 million tons of molten iron produced in Korea in 2018 were reduced using hydrogen and considering the endothermic reaction caused by reduction, a total of 17.4 TWh of energy would be required per month. In other words, an enormous amount of energy equivalent to about 40% of Korea's total electricity demand would be needed. This is a matter that requires serious consideration at the national or global level, beyond the capabilities of the steel industry alone. Currently, based on existing technology, nuclear power including the nuclear fusion is the least cost-intensive method for hydrogen production. It is also crucial to actively explore the utilization of various models of Small Modular Reactors (SMRs), which are being developed in various countries, for the hydrogen reduction process.[75] Since hydrogen reduction steelmaking is challenging to implement within a single steel company, major steelmakers are promoting related technology development as a national project.

As a state-supported large-scale project, the EU has undertaken the ULCOS project,[76] while Japan has initiated the COURSE-50 project.[77] The ULCOS project aimed to achieve a 50% reduction in CO_2 but was discontinued at the demo plant stage. Sweden's SSAB is currently advancing the HYBRIT project (Hydrogen BReakthrough Ironmaking Technology) with EU funding and aims to commercialize it by 2026.[78] Primetals, a steel engineering company, is piloting the HYFOR (HYdrogen-Based Fine Ore Reduction) process at VAS in Austria.[79] The COURSE-50 project involves the participation of all Japanese steel companies and related engineering firms to achieve a 30% reduction in CO_2. In Korea, a national project is underway to achieve a 10% CO_2 reduction, and the second phase of the project, which aims to strengthen the target, has recently been launched.[80]

Considering the significant costs and time involved in adopting full-scale hydrogen reduction in the ironmaking process, efforts are being made to reduce CO_2 emissions by injecting COG (Coke Oven Gas, with a hydrogen content of 50–60% and CH_4 content of 20–30%) into the BF as an intermediate step.[81] In the hydrogen reduction process of iron ore, finer ore particle sizes result in faster reduction rates, so the use of fine particle size ore as raw material is being explored.

Among the new ironmaking processes listed in Table 8.1, the MIDREX process utilizes LNG as an ore heating and reducing agent. By incorporating hydrogen into LNG, further reductions in CO_2 emissions can be achieved. In the case of FINEX, as shown in Figure 8.2, the gas entering the FBR already contains more than 10% hydrogen. The addition of LNG or hydrogen gas at this stage can help reduce CO_2 generation. Based on these considerations, two hydrogen reduction processes can be mentioned for full decarbonization.

(A) MIDREX process: Very fine ore (<40 μm) → Pellet → MIDREX → Reduced iron → EAF (Electric Arc Furnace) → Steel refining process

(B) FINEX process: Fine ore (<8 mm) → FINEX fluidized bed reactor → Reduced iron → ESF (Electric Smelting Furnace) → Converter → Steel refining furnace

Important differences between the two processes are the type of iron ore in terms of the cost and the availability, and the melting facilities of the reduced iron either in ESF + Convertor (B) or in EAF (A). Since the fine ore used in (B) is cheaper and more widely available than the very fine ore used in (A), FINEX has the advantage. The comparison of the two melting processes is very complicated when variables such as iron recovery, reduced iron, slag processing, heat supply, melting equipment, and final steel products are considered, but the steel production costs appeared to be quite comparable although they are much higher than those in the current process. Therefore, the success of the technology development lies in the way to save production costs. The total competitiveness is yet to be seen until the complete commercialization of the two processes.

It is fortunate to see that zero carbon steel reduction is theoretically possible with the ore reduction processes by hydrogen currently under development by major steel makers worldwide. However, in addition to the hydrogen reduction process, the steelmaking process is to be newly developed for the production of high-quality steel products equivalent to the current level. Numerous changes to steelworks facilities are necessary, along with the need for additional green energy for hydrogen production and heating. Consequently, the manufacturing cost of steel inevitably increases, as mentioned earlier. However, carbon reduction is not an option, but an essential for the future of human life. The biggest cost burden will come from hydrogen production using electricity generated by renewable energy. In this regard, close cooperation among steelmakers not only with steel business societies but with non-steel business societies is strongly recommended. Also, the steel industry should not solely focus on the hydrogen reduction process, but explore options such as CCS (Carbon Capture and Storage) and initiatives like phytoplankton farming to enhance CO_2 absorption capacity in the ocean, or Cquestrate[82] which utilizes lime to absorb CO_2.

Reducing iron ore with green hydrogen means that iron is produced with a large consumption of electric energy instead of coal. As a result, various processes for producing iron using electricity have been proposed and attracted some attention from the industry. Representative examples are Professor Sadoway's Molten Oxide Electrolysis (MOE) and ULCOS's ULCOWIN.[83, 84] MOE is to use electricity to directly reduce molten iron oxides by electrolysis into metallic iron form, whereas ULCOWIN is a low-temperature electrolysis process that produces solid-state elemental iron from iron oxides. Arcelor Mittal is interested in the commercialization of these two processes with EU financial support. However, it will need some time to conclude whether any of the technologies is industrially feasible.[85, 86]

8.4 STEEL INDUSTRY IN THE FOURTH INDUSTRIAL REVOLUTION – SMART FACTORY

Recently, there has been a significant focus on the concept of the Fourth Industrial Revolution, which is anticipated to revolutionize the future industrial system.[87] In the context of manufacturing, the Fourth Industrial Revolution manifests as the establishment of the intelligent factory, known as the "Smart Factory." A smart factory is a system that leverages artificial intelligence (AI), big data, and the Internet of Things (IoT) to manufacture products, facilitate sales, and engage in quality sharing with customers.[88, 89] The impact and implications of smart factory implementation on various processes and sales are still under discussion, and new concepts continue to emerge. Therefore, it is crucial to proactively analyze the influence of smart factories on the future of the steel industry.

8.4.1 Fourth Industrial Revolution and Smart Factory

From a forward-looking perspective, the significance of the Fourth Industrial Revolution is being emphasized across all sectors. While there is no universally established definition for the Fourth Industrial Revolution, we define it as a driving force that generates new value in manufacturing. This new value can be created not only by fostering innovation within existing industries but also by establishing entirely new industries through the effective utilization of information technology (IT) knowledge.[87] In the realm of manufacturing, terms such as smart factory, digital transformation, and digitalization are employed to realize the objectives of the Fourth Industrial Revolution. Since these terms share the same ultimate goal, we will use the term smart factory here to represent them collectively.

Germany, at the forefront of promoting the Fourth Industrial Revolution, introduced the concept of "Industry 4.0" as a national policy in the manufacturing industry. They implemented an intelligent production system that incorporated IT knowledge across the entire manufacturing value chain and defined it as a smart factory.[88, 89] Based on the aforementioned definition of the Fourth Industrial Revolution, a smart factory aims to enhance competitiveness and achieve sustainable growth in the steel industry by leveraging new IT technologies to improve manufacturing efficiency and capabilities, following the same principles that apply to other sectors.

In the manufacturing industry, IT knowledge has been utilized for a considerable period, and automation stands as a prime example. Within the steel industry, automation has been identified as a pressing requirement due to the challenges posed by high temperatures, heavy loads, and fast-paced operations. With cutting-edge IT technology, the immense machines in factories have been automated, reducing the need for human workers on the shop floor. The automation of production facilities can lead to improvements in productivity, consistent quality, and enhanced cost competitiveness by minimizing reliance on manual labor.

The smart factory holds great promise as a means of creating new levels of innovation in the manufacturing industry, thereby generating additional value for humans. Advanced industrialized nations view this as an opportunity to widen the gap further with latecomers through the establishment of smart factory systems. Conversely, latecomers see it as a golden opportunity to catch up with advanced countries swiftly by building new smart factory systems from the ground up.

When it comes to automation, what sets a smart factory apart from a conventional factory? The answer lies in the term !smart! itself. Traditional automation follows predetermined sequences and procedures. In contrast, a smart factory employs artificial intelligence (AI) to continuously sense, analyze, and learn from phenomena and operating conditions during production. It seeks to identify optimal operating conditions and performs intelligent control of equipment. Various approaches, such as big data-based reinforcement learning methods, are being employed to determine the optimal conditions for operation.[90]

Big data possesses three key characteristics, known as the 3V: volume, velocity, and variety. It typically encompasses terabytes or even larger capacities and can include unstructured data such as images and sounds. Various techniques are being developed to enhance the utilization of big data, ranging from data storage and distribution to collection and analysis processing. AI, on the other hand, refers to a system in which computers exhibit human-like functions such as learning, reasoning, adaptation, and argumentation. Through machine learning and deep learning algorithms that continuously learn from data, AI can uncover hidden insights and create optimal conditions for itself. Thus, a machine actively controlled by AI is referred to as a smart machine. With repeated learning, these

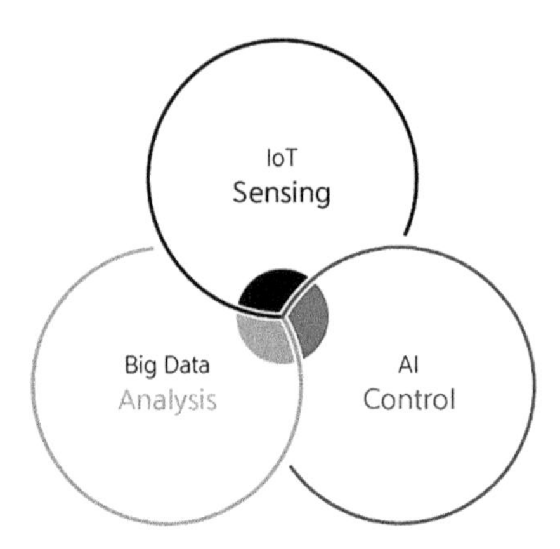

FIGURE 8.11 New IT technology and basic concept of the smart factory.[90]

machines become increasingly intelligent, eventually surpassing the human possessing capabilities.[91]

In terms of increasing the competitiveness and sustainability of the manufacturing industry, including the steel industry, there is a consensus on the importance of three key IT technologies: big data, AI, and IoT. Figure 8.11 illustrates the fundamental concept of a smart factory, where these three technologies play a central role.[90]

IoT plays a vital role in remotely collecting data from various sensors attached to facilities. Real-time data, including video, is collected rapidly from all operating facilities, resulting in large capacities of data. Hence, the data associated with a smart factory is referred to as big data. AI analyzes this big data and possesses the capability for intelligent control to optimize the operation of equipment.

The objective of a smart factory is to transform into a more efficient environment compared to the current level of production operations. Achieving this requires the seamless integration of advanced IT knowledge with specialized domains that apply smart factory concepts. Figure 8.12 presents a schematic representation of the concept of convergence.[90] The specific field where the smart factory is implemented is known as the domain, and in this case, it is the steel industry. To successfully build a smart factory, cutting-edge IT expertise must be harmoniously combined with domain-specific knowledge in the steel industry. During this convergence, experts in the steel industry should take the lead. If IT experts become the primary drivers without sufficient expertise in smart factory value creation, the resulting system may become biased towards IT knowledge, potentially deviating from the goal of optimizing the process.

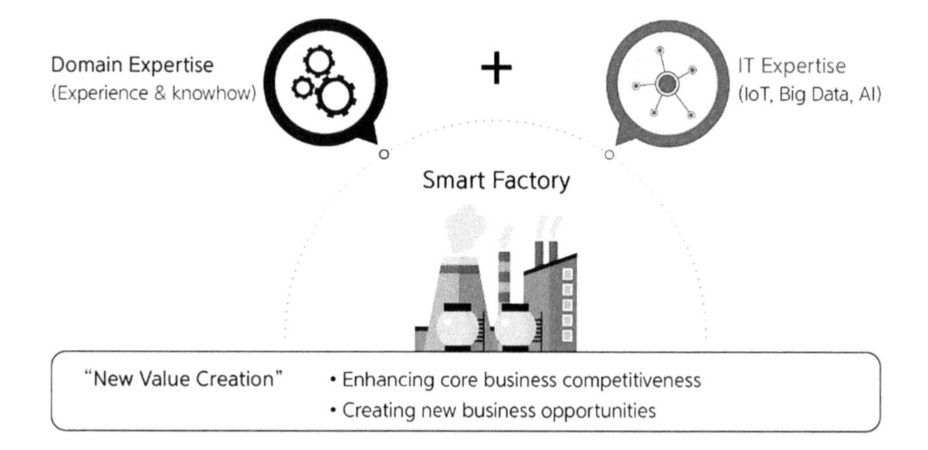

FIGURE 8.12 Concept of convergence for value creation in smart factory.[90]

8.4.2 Smart Factory in the Steel Industry

A platform is essential to enhance the effectiveness of smart factories, enabling field technicians without extensive IT knowledge to operate the factory intelligently according to their requirements. Figure 8.13 illustrates the schematic structure of PosFrame, a smart factory platform implemented in POSCO's integrated steel process.[92] This platform possesses the capability to sense and analyze big data, derive optimal values, and control facilities based on these values. Furthermore, it offers a common software layer, including an application protocol interface, which supports various application software operating on the platform in an

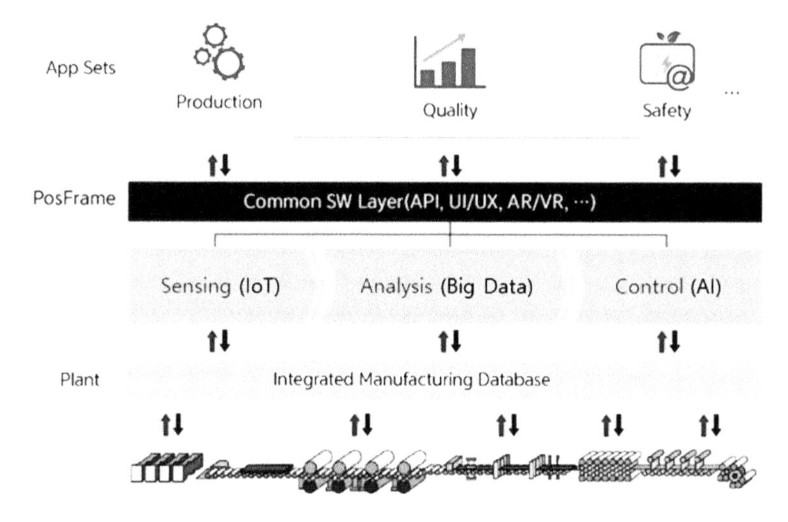

FIGURE 8.13 Outline of PosFrame, the POSCO smart factory platform.[92]

app-type format. The platform also incorporates other essential functions, such as user interface/user experience and augmented reality/virtual reality, within this layer.

When field technician requires specific data, they can easily access it from the virtual database on the platform screen with just a mouse click, eliminating the need for a direct sensor connection. Additionally, the platform includes pre-installed big data and AI software necessary for smart factories, enabling field technicians to select desired algorithms and conduct the required analysis and control. Prominent platforms in the industry include GE Predix(93) and Siemens' MindSphere,(94) with other companies such as ABB,(95) Bosch,(96) and SAP(97) also developing their smart factory platforms.

The steel industry exhibited an early interest in AI compared to other industries. Active efforts were made to automate operations by introducing the expert system, an AI technique that gained popularity in the 1990s, into the steel manufacturing process. During that time, steelmakers compiled the knowledge of field experts into an expert system database, established rules for equipment control, and stored them in computers.(98) However, over time, the usefulness of this approach diminished as there was a lack of continuous efforts to update the database. Subsequently, interest in AI waned. However, after AlphaGo's overwhelming victory over Lee Se-dol in the Go match in Korea in 2016,(99) interest in AI resurfaced, and the active promotion of smart factory construction led to the emergence of many successful cases.

Most smart factories in the steel industry are currently being developed at the company level, generating significant interest. Since 2016, POSCO has been at the forefront of building smart factories and has achieved notable advancements across all areas of its business site. Examples of applications include IoT-based smart safety helmets, big data and AI-driven BF operation control, optimization of steelmaking converter and RH operations, plate shaping, CGL coating thickness control, and power plant NO_x generation control.(100) In Germany, Siemens is spearheading the development and construction of smart factories, also known as digitalization or Industry 4.0. Steelmakers such as AM,(101) TKS,(102) and TSE(103) are actively establishing smart factories with a focus on AI utilization.

SMS, a major German steel engineering company, announced its partnership with Noodle.ai, an AI company, to supply process optimization solutions to Big River Steel, an American steelmaker.[104] Additionally, SSAB America has been implementing AI technology to enhance production efficiency.[105] In the United States, Nucor is also actively promoting digitalization to improve safety measures and enhance customer engagement.[106] In Japan, both NSC[107] and JFE[108] primarily integrate AI techniques into their existing computer models to enhance optimization capabilities. NSC is working towards establishing a smart factory across the entire steelmaking process and has recently developed a smart safety helmet called "Anzen Mimamori-kun."[107]

In China, the concept of Germany's Industry 4.0 has been widely adopted across various industries, with national-level activities taking place.[109] Baowu, a state-owned steel company, is leading the Industry 4.0 initiatives in the Chinese steel industry, and several other steel companies, such as Shagang and Yuzhou Steel, are also participating.[110] Baowu has designated "HSM 1580" as a pilot plant and aims to increase production by over 15% while reducing inventory by 50%.[111] The concept of a smart factory is not limited to steel processing in China, as existing e-commerce companies are also incorporating it. Even Ouyeel is actively working towards convergence.[112]

When it comes to building a smart factory, it is efficient to initially utilize AI in individual processes before developing a comprehensive platform. This approach is more feasible as it requires less investment and the acquisition of experts compared to attempting to establish a complete platform right from the start. By analyzing existing data and applying AI through cost-effective software tailored to specific process improvements, it becomes possible to generate high-added value with minimal effort. Once the potential for value creation is realized, expansion into other areas can be pursued. As the need for AI analysis expands to encompass various types and larger volumes of data, the step-by-step implementation of IoT and big data infrastructure can be undertaken.

In the case of integrated steelworks with a well-equipped automation system, various optimization techniques can be applied in production management, process design, process control, and quality control. Optimization is crucial for the automation process, and techniques like the minimum gradient method are utilized for the automatic control of processes. By incorporating AI techniques in these areas, a wide range of applications becomes possible. For instance, by analyzing the correlation between vibration data from a rolling mill and changes in the thickness or profile of the steel sheet, optimal operating conditions for the mill can be derived. This analysis can lead to improvements in sheet quality and a reduction in defect rates.

Smart factories can also benefit customer service in the steel industry. Providing customers with image data depicting changes in mechanical properties of steel sheets in length, width, and surface conditions, including location information, enables them to enhance the efficiency of post-processing steel products. Efficiency can be increased by identifying and removing parts that do not meet specifications or optimizing the processing method to match specific characteristics. Surface defects spanning the entire length of the coil can be transformed into image-based digital data, forming part of big data. By establishing a smart factory through the digitalization of such vast data, both steelmakers and customers can access a wider range of innovations.

The smart factory is anticipated to find applications not only in manufacturing but also across various fields, including management, finance, healthcare, welfare, and culture. Consequently, when applied to the field of management, it can be referred to as smart management, while in finance, it can be called smart financing. PwC, a global management consulting firm, has estimated that AI has the potential to increase global GDP by US$15.7 trillion by 2030.[(113)] Recently, the ChatGPT has been released, allowing the general public to obtain the necessary information by engaging in natural language conversations with AI. It is predicted that the scale, scope, and complexity of AI-driven innovation will be unprecedented in human history.

REFERENCES

1. Y. H. Lee, Internal report from POSRI, 2021.
2. Valentin Vogl, Olle Olsson, and Björn Nykvist, “Phasing out the blast furnace to meet global climate targets”, *Joule*, Vol. 5, October 20, 2021, pp. 2646–2662, Published by Elsevier Inc., https://doi.org/10.1016/j.joule.2021.09.007
3. Nobuo Kato, “Analysis of structure of energy consumption and dynamics of emission of atmospheric species related to the global environmental change: SOx, NOx, and CO2) in Asia”, *Atmos. Environ.*, Vol. 30, No. 5, 1996, pp. 757–785
4. Min Gan, Xiaohui Fan, Xuling Chen, Zhiyun Ji, Wei Lv, Yi WANG, Zhiyuan Yu, and Tao Jiang, “Reduction of pollutant emission in iron ore sintering process by applying biomass fuels”, *ISIJ Int.*, Vol. 52, No. 9, 2012, pp. 1574–157, https://www.jstage.jst.go.jp/article/isijinternational/52/9/52_1574/_pdf

5. Takao Harada, Osamu Tsuge, Isao Kobayashi, Hidetoshi Tanaka, and Hiroshi Uemura, "The development of new iron making processes", *Kobelco Technolo. Rev.*, No. 26, December 2005, p. 9, https://www.kobelco.co.jp/english/ktr/pdf/ktr_26/092-097.pdf
6. "COREX® Efficient and environmentally friendly smelting reduction", *Primetals*, https://www.primetals.com/fileadmin/user_upload/content/01_portfolio/1_ironmaking/corex/COREX.pdf
7. N. J. Goodman, "The HIsmelt technology from Australia to China....and Back Again?" *Acadia Iron*, https://az659834.vo.msecnd.net/eventsairseasia-prod/production-ausimm-public/a4da4335f20d4bbaacc2e914cb3e1734
8. "HISARNA, building a sustainable steel industry", https://www.tatasteel-europe.com/sites/default/files/tata-steel-europe-factsheet-hisarna.pdf
9. "THE FINEX®, process economical and environmentally safe iron making", *POSCO and Primetals*, https://www.primetals.com/fileadmin/user_upload/content/01_portfolio/1_ironmaking/finex/THE_FINEX_R__PROCESS.pdf
10. Thomas Battle, Urvashi Srivastava, John Kopfle, Robert Hunter, and James McClelland, "The direct reduction of iron", *Treatise on Process Metallurgy*, Vol 3, 2014, Elsevier Ltd, http://dx.doi.org/10.1016/B978-0-08-096988-6.00016-X
11. H. Y. L. Tenova, "A pioneer in direct reduction, designing and supplying advanced DR plants", https://tenova.com/about-us/our-brands/tenova-hyl
12. S. H. Lee, POSCO, Private Communication
13. "MIDREX® process", https://www.midrex.com/technology/midrex-process/
14. M. Atsushi, H. Uemura, and T. Sakaguchi, "MIDREX® processes", *Kobelco Technolo. Rev.*, No. December 29, 2010, pp. 50–57, https://www.kobelco.co.jp/english/ktr/pdf/ktr_29/050-057.pdf
15. "Arvedi ESP — real endless strip production", *Primetals*, https://www.primetals.com/portfolio/endless-casting-rolling/arvedi-esp
16. S. H. Lee, "A casting speed point of view in POSCO", *BHM Berg- und Hüttenmännische Monatshefte*, Vol. 163, No. 1, January 2018, pp. 3–10, doi:10.1007/s00501-017-0695-3
17. "Rizhao steel produces first coil on fifth Arvedi ESP line from Primetals Technologies", Primetals, London, January 14, 2021, https://www.primetals.com/fileadmin/user_upload/press-releases/2021/20210114/PR2021012211en.pdf
18. "United States steel corporation breaks ground on the most technologically advanced steel mill in North America", *US Steel*, February 9, 2022, https://investors.ussteel.com/news/news-details/2022/United-States-Steel-Corporation-Breaks-Ground-on-the-Most-Technologically-Advanced-Steel-Mill-in-North-America/default.aspx
19. G. E. Sa, I. S. A. C. Mihaiela, and Roderick Ian Lawrence Guthrie, "Progress of strip casting technology for steel; Historical developments", *ISIJ Int.*, Vol. 52, No. 12, 2012, pp. 2109–2122, https://www.jstage.jst.go.jp/article/isijinternational/52/12/52_2109/_pdf

20. D. J. Sosinsky, P. Campbell, Rama Mahapatra, W. Blejde, and F. Fisher, "The CASTRIP ® proc ess - Recent developments at Nucor steel's commercial strip casting plant", *Metallurgist*, Vol. 52, 2009, pp. 691–699, doi:10.1007/s11015-009-9116-5. https://www.researchgate.net/publication/227305690_The_CASTRIP_R_process_-_Recent_developments_at_Nucor_steel%27s_commercial_strip_casting_plant
21. "China's first! Shagang achieved industrial production of Castrip", http://eng.shasteel.cn/doc/2019/05/15/10171.shtml
22. Y. J. Jung, "POSCO Wins 'innovation of the year' award from world steel association", October 15, 2015, http://www.koreaittimes.com/news/articleView.html?idxno=54573
23. R. Hadfield, "A research on Faraday's 'steel and alloys", *Philosophical Transactions of the Royal Society of London, Series A*, Vol. 230, 221–292. [Plates 4-12.], Printed and Published for the Royal Society, 1931.
24. R. G. J. Edyvean and C. Hammond, "The metallurgical work of Henry Clifton Sorby and an annotated catalogue of his extant metallurgical specimens", *Historical Metallurgy*, Vol. 31, No. 2, 1997, https://www.google.com/url?sa=t&rct=j&q=&esrc=s&source=web&cd=&cad=rja&uact=8&ved=2ahUKEwi4ueGm1a6BAxWqm1YBHZIGCQEQFnoECBAQAQ&url=https%3A%2F%2Fwww.hmsjournal.org%2Findex.php%2Fhome%2Farticle%2Fdownload%2F317%2F306&usg=AOvVaw2cpj1udlZyOZJKf1OY6Bsh&opi=89978449
25. G. Tweedale and W. D. MacDonald Paton, "Sir Robert Abbott Hadfield F. R. S. (1858–1940) and the discovery of manganese steel", *The Royal Society Publishing*, November 1, 1985, https://doi.org/10.1098/rsnr.1985.0004
26. Javier Yanes, "Stainless steel, the most sustainable material", *bbvaopen mind.com*, August 11, 2021, https://www.bbvaopenmind.com/en/technology/visionaries/harry-brearley-stainless-steel-sustainable-material/
27. "Infographic - Fit for 55: Why the EU is toughening CO_2 emission standards for cars and vans, Council of the EU and the European Council", December 12, 2022, https://www.consilium.europa.eu/en/infographics/fit-for-55-emissions-cars-and-vans/
28. "Automotive steel, POSCO", https://www.google.co.kr/url?sa=t&rct=j&q=&esrc=s&source=web&cd=&ved=2ahUKEwiq-6r4kPH8AhXGp1YBHV_kCNMQFnoECBUQAQ&url=https%3A%2F%2Fproduct.posco.com%3A4451%2Fhomepage%2Fproduct%2Fcommon%2Fs91pdownload.jsp%3Ffile%3D%2Fhfiles%2Fproductpr%2F779da3fa1768ec9fecfe40485840566d.pdf%26filename%3D2021%25B3%25E2%2520%25C0%25DA%25B5%25BF%25C2%25F7%25B0%25AD%25C6%25C7%2520%25C4%25AB%25C5%25BB%25B7%25CE%25B1%25D7_%25BF%25B5%25B9%25AE_Rev.1.pdf&usg=AOvVaw0k-UlOxykqB3iz89ZKHjhE
29. Xiaohua Hu and Zhili Feng, "Advanced high-strength steel—basics and applications in the automotive industry", April 2021, Oak Ridge National Laboratory, ORNL/TM-2021/2047, https://info.ornl.gov/sites/publications/Files/Pub158668.pdf

30. "[POSCO Product] POSCO's 5 types of high manganese steel", https://www.youtube.com/watch?v=1Nkb_5YSXeQ
31. B. C. De Cooman, K. G. Chin, and J. K. Kim, "High Mn TWIP steels for automotive applications", https://cdn.intechopen.com/pdfs/13349/InTech-High_mn_twip_steels_for_automotive_applications.pdf
32. S. K. Kim, J. Choi, S. C. Kang, I. R. Shon, and K. G. Chin, ""Development of high strength TWIP steels", *POSCO Techn. Rep.*, 2006,vol. 10, no. 1, 106.
33. Y. G. Kim, T. W. Kim, and S. B. Hong, "High strength formable automotive structure steel", Proceedings of ISATA, Aachen, Germany, September 1993, p. 269
34. Iung Thierry, Petitgand Gerard, and Staudte Jonas, "Method for producing a Twip steel sheet having an austenitic microstructure", WIPO Patent Application WO/2017/203343, ArcelorMittal, https://www.sumobrain.com/patents/wipo/Method-producing-twip-steel-sheet/WO2017203343A1.html
35. C. Scott, S. Allain, M. Faral, and N. Guelton, "The development of a new Fe-Mn-C austenitic steel for automotive applications", *Rev. Metall-Cah. Inf. Tech.*, Vol. 103, 2006, pp. 293–302, doi:10.1051/metal:2006142, https://www.researchgate.net/publication/248865027_The_Development_of_a_New_Fe-Mn-C_Austenitic_Steel_for_Automotive_Applications/citation/download
36. Ping Bao, "Baosteel new energy vehicles comprehensive material solutions", April 2021, https://res.baowugroup.com/attach/2021/04/28/6ca3c79613a04dd192667c2f1801fbb1.pdf
37. "Mangalloy", last edited January 19, 2023, https://en.wikipedia.org/wiki/Mangalloy
38. J. W. Moon, "POSCO to produce & supply high manganese steel to ExxonMobil for slurry pipes", March 22, 2017, https://www.mk.co.kr/economy/view.php?sc=50000001&year=2017&no=194915
39. Maurice Smith, "ExxonMobil adopts manganese steel pipe for Kearl Oilsands mining operations", Thursday, March 23, 2017, https://www.jwnenergy.com/article/2017/3/23/exxonmobil-adopts-manganese-steel-pipe-kearl-oilsa/
40. "POSCO lays the basis for expanding sales of the world's first mass-produced cryogenic high manganese steel", June 8, 2022, https://newsroom.posco.com/en/posco-lays-the-basis-for-expanding-sales-of-the-worlds-first-mass-produced-cryogenic-high-manganese-steel/
41. Naida Hakirevic Prevljak, "DSME installs world's 1st high-manganese steel LNG fuel tanks on VLCC", June 16, 2022, https://www.offshore-energy.biz/dsme-installs-worlds-1st-high-manganese-steel-lng-fuel-tanks-on-vlcc/
42. "Suitability of high manganese austenitic steel for cryogenic service and development of any necessary amendments to the IGC code and IGF code, Technical information regarding high manganese austenitic steel for cryogenic service", Submitted by the Republic of Korea Sub-committee on carriage of cargoes and containers, 3rd session, Agenda item 8, CCC 3/8, June 3, 2016, IMO

43. "Anti perforation warranty", https://www.hyundai.co.uk/owning/5_year_warranty/anti-perforation__warranty
44. "Galvanized steel, POSCO", http://www.steel-n.com/e-sales/pdf/en/galvanizedsteel.pdf
45. S. Schuerz, M. Fleischanderl, G. H. Luckeneder, K. Preis, T. Haunschmied, G. Mori, and A. C. Kneissl, "Corrosion behaviour of Zn–Al–Mg coated steel sheet in sodium chloride-containing environment", *Corros. Sci.*, Vol. 51, No. 10, 2009, pp. 2355–2363, ISSN 0010-938X, https://doi.org/10.1016/j.corsci.2009.06.019
46. "PosMAC® POSCO magnesium aluminium alloy coating product", https://www.google.co.kr/url?sa=t&rct=j&q=&esrc=s&source=web&cd=&ved=2ahUKEwiL0ea4vPP8AhWhglYBHTheBFAQFnoECAwQAQ&url=https%3A%2F%2Fwww.poscoindonesia.id%2Ffiledown.jsp%3Ffilename%3D200000_A1__PosMAC_Catalog.pdf&usg=AOvVaw070p3NwQ66w5NxixtqkLHO
47. R. Sohn, T. C. Kim, G. I. Ju, M. S. Kim, and J. S. Kim, "Anti-corrosion performance and applications of PosMAC® steel", *Corros. Sci. Technol.*, Vol. 20, No. 1, 2021, pp. 7–14, https://koreascience.kr/article/JAKO202116056968039.pdf
48. POSCO Internal Report
49. "Development of POSCO PosMAC® super", *Steel Daily*, Vol. 12, 2021, p. 28 (in Korean), https://www.steeldaily.co.kr/news/articleView.html?idxno=160724
50. "Jet Vapor Deposition – A ground-breaking steel coating technology", https://corporate.arcelormittal.com/media/case-studies/jet-vapor-deposition-a-ground-breaking-steel-coating-technology
51. T. Y. Kim and M. Goodenough, "Simultaneous Co-deposition of Zn-Mg alloy layers on steel strip by PVD process". *Corros. Sci. Technol.*, Vol. 10, No. 6, 2011, pp. 194–198. https://doi.org/10.14773/CST.2011.10.6.194
52. J. W. Kang, J. M. Park, S. H. Hwang, S. H. Lee, K. M. Moon, and M. H. Lee, "Influence of heat treatment and magnesium content on corrosion resistance of Al-Mg coated steel sheet", *J. Korean Inst. Surf. Eng.*, Vol. 49, No. 2, 2016, http://dx.doi.org/10.5695
53. J. W. Park, K. H. Lee, B. K. Park, and S. H. Hong, "Development of anticorrosive coating technique for alloy plated steel sheet using silane based organic-inorganic hybrid materials", *Corros. Sci. Technol.*, Vol. 12, No. 6, 2013, pp. 295–303, https://doi.org/10.14773/CST.2013.12.6.295
54. D. S. Hwang, D. C. Cho, H. J. Yoo, J. S. Kim, and I. W. Cheong, "Parametric study on gloss property of UV curable coated steel", *J. Adhes. Interf.*, Vol. 15, No. 3, 2014, https://koreascience.kr/article/JAKO201435053629669.pdf
55. D. H. Cho, J. R. Lee, S. G. Noh, and J. T. Kim, "Excellent heat-dissipating black resin composition, a method for treating a steel sheet using the same and steel sheet treated thereby", KR100804934B1 (in Korean), https://patents.google.com/patent/KR100804934B1/en
56. "PosART, POSCO Steeleon", https://posart.poscosteeleon.com/kr/main.do
57. "Stainless steel anti-bacterial material", https://www.bsstainless.com/stainless-steel-anti-bacterial-material

58. K. P. Ko, N. B. Park, J. S. Kim, and H. H. Lee, "Composition for organic-inorganic hybrid coating solution and steel sheet having organic-inorganic hybrid coating", KR101449109B1 (in Korean), https://patents.google.com/patent/KR101449109B1/en
59. T. C. Kim, S. H. Kim, M. S. Oh, H. H. Yu, and J. S. Kim, "Surface treatment method for high strength steel and plating method the same", KR20160106528A (in Korean), https://patents.google.com/patent/KR20160106528A/en
60. M. H. Hong, I. K. Kim, and H. J. Yun, "Effect of inorganic filler addition on non-combustible and mechanical properties of color coated steel sheets" (in Korean) *J. Met. Mater.*, Vol. 58, No. 11, 2020, pp. 1–8, http://www.kjmm.org/upload/pdf/kjmm-691.pdf
61. Tony Seba, "Clean disruption of energy and transportation", May 20, 2014, https://tonyseba.com/wp-content/uploads/2014/05/book-cover-Clean-Disruption.pdf
62. "Latest developments in steelmaking capacity", https://www.oecd.org/industry/ind/latest-developments-in-steelmaking-capacity-2022.pdf
63. Rod Beddows, "Steel 2050: How steel transformed the world and now must transform itself", *Paperback*, October 6, 2014, https://www.amazon.com/Steel-2050-Transformed-Transform-Itself/dp/0993038107
64. "Accenture strategy, steel demand beyond 2030 forecast scenarios", Presented to OECD, Paris, September 28, 2017, https://www.oecd.org/industry/ind/Item_4b_Accenture_Timothy_van_Audenaerde.pdf
65. "Scrap use in the steel industry", Fact sheet from World Steel Association
66. "Steelmaker Nucor looks to automotive sector for growth", October 22, 2020, https://www.spglobal.com/marketintelligence/en/news-insights/latest-news-headlines/steelmaker-nucor-looks-to-automotive-sector-for-growth-60879355
67. "HYBRIT. A new revolutionary steelmaking technology", https://www.ssab.com/en/fossil-free-steel/hybrit-a-new-revolutionary-steelmaking-technology
68. "Hydrogen-based steelmaking to begin in Hamburg", https://corporate.arcelormittal.com/media/case-studies/hydrogen-based-steelmaking-to-begin-in-hamburg
69. "The future of steel — hydrogen-based steelmaking", https://newsroom.posco.com/en/exploring-hydrogen-with-posco-3-the-future-of-steel-hydrogen-based-steelmaking/
70. "Promotion of innovative technology development", https://www.nippon-steel.com/en/csr/env/warming/future.html
71. "Baowu group starts construction on hydrogen shaft furnace project", February 24, 2022, https://www.steelorbis.com/steel-news/latest-news/baowu-group-starts-construction-on-hydrogen-shaft-furnace-project-1234715.htm
72. S. H. Yi, W. J. Lee, Y. S. Lee, and W. H. Kim, "Hydrogen-based reduction ironmaking process and conversion technology", *Korean J. Met. Mater.*, Vol. 59, No. 1, 2021, pp. 41–53, doi:10.3365/KJMM.2021.59.1.41, http://kjmm.org/upload/pdf/kjmm-2021-59-1-41.pdf

73. M. Wanner, “Transformation of electrical energy into hydrogen and its storage”, *Eur. Phys. J. Plus*, Vol. 136, 2021, p. 593, https://doi.org/10.1140/epjp/s13360-021-01585-8
74. “Settlement unit price by fuel source, electricity market statistics”, *Korea Power Exchange*, July 13, 2022 (in Korean), https://kosis.kr/statHtml/statHtml.do?orgId=388&tblId=TX_38804_A016&conn_path=I2
75. “Small modular reactors”, https://www.iaea.org/topics/small-modular-reactors
76. “Ultra-Low CO_2 steelmaking, fact sheet”, https://cordis.europa.eu/project/id/515960
77. “To the future of the low carbon blast furnace CO_2 ultimate reduction system for cool earth 50 (COURSE50) project”, https://www.course50.com/en/
78. “HYBRIT receives support from the EU Innovation Fund”, *Press Release, Hydrogen*, April 1, 2022, https://group.vattenfall.com/press-and-media/pressreleases/2022/hybrit-receives-support-from-the-eu-innovation-fund
79. “HYFOR pilot plant under operation – The next step for carbon free, hydrogen-based direct reduction is done”, London, June 24, 2021, https://www.primetals.com/press-media/news/hyfor-pilot-plant-under-operation-the-next-step-for-carbon-free-hydrogen-based-direct-reduction-is-done
80. “Preliminary passage of the decarbonization steel technology development project”, *Iron and Steel Newspaper*, Vol. 11, 2022, p. 1 (in Korean), https://www.snmnews.com/news/articleView.html?idxno=507090
81. “Paul Wurth books new order for coke oven gas compression and injection technology at HKM”, September 8, 2020, https://www.paulwurth.com/en/paul-wurth-books-new-order-for-coke-oven-gas-compression-and-injection-technology-at-hkm/
82. Chris Unitt, “Cquestrate – A crowdsourced solution to climate change”, July 29, 2008, https://www.chrisunitt.co.uk/2008/07/cquestrate-a-crowd-sourced-solution-to-climate-change/
83. Elizabeth A. Thomson, “Engineers forge greener path to iron production”, *MIT News*, August 25, 2006, https://news.mit.edu/2006/iron
84. K. West, “Low-temperature Electrowinning for steelmaking (ULCOWIN)”, *Technology Factsheet*, https://energy.nl/wp-content/uploads/ulcowin-technology-factsheet_080920-7.pdf
85. “ArcelorMittal invests $36 million in steel decarbonisation disruptor Boston Metal”, January 27, 2023, https://corporate.arcelormittal.com/media/press-releases/arcelormittal-invests-36-million-in-steel-decarbonisation-disruptor-boston-metal
86. S. Koutsoupa, S. Koutalidi, and Evangelos Bourbos, “Electrolytic iron production from alkaline bauxite residue slurries at low temperatures carbon-free electrochemical process for the production of metallic iron”, *Johnson Matthey Technol. Rev.*, Vol. 65, No. 3, 2021, pp. 366–374, https://doi.org/10.1595/205651320X15918757312944
87. “Fourth industrial revolution”, https://www.weforum.org/focus/fourth-industrial-revolution

88. "What is a smart factory?", https://www2.deloitte.com/us/en/pages/consulting/solutions/the-smart-factory.html/#info3
89. "Germany: Industrie 4.0", January 2017, https://ati.ec.europa.eu/sites/default/files/2020-06/DTM_Industrie%204.0_DE.pdf
90. D. H. Choi, "Manage with Smart Factory", Seoul: Huckleberry Books, 2019 (in Korean)
91. "Reinforcement learning 101", https://towardsdatascience.com/reinforcement-learning-101-e24b50e1d292
92. "PosFrame", *POSCO DX*, https://www.poscodx.com/eng
93. "Predix platform", https://www.ge.com/digital/iiot-platform
94. "MindSphere", https://siemens.mindsphere.io/en
95. "ABB Ability™", https://global.abb/topic/ability/en
96. "Nexeed — Welcome to the smart factory", https://www.bosch.com/stories/nexeed-smart-factory/
97. "Smart factory solutions", https://atos.net/en/lp/sap_smartfactory
98. J. Dorn, "Expert systems in the steel industry", *IEEE Expert*, Vol. 11, No. 1, February 1996, p. 18, doi:10.1109/MEX.1996.482952, https://ieeexplore.ieee.org/document/482952
99. "The challenge match", https://www.deepmind.com/research/highlighted-research/alphago/the-challenge-match
100. "POSCO the lighthouse factory #1: POSCO's smart factory shines light on manufacturing industry (the concept)", https://newsroom.posco.com/en/posco-lighthouse-factory-1/
101. "Industry 4.0", https://corporate.arcelormittal.com/media/case-studies/industry-4-0
102. "Intelligent and connected processes, smart factory", https://www.thyssenkrupp-steel.com/en/company/digitalization/smart-factory/smart-factory.html
103. "Smarter technologies and services for a sustainable automotive value chain", https://www.tatasteeleurope.com/automotive/industry-themes/digitalisation
104. "Big river steel's flat steel complex on its way to become a learning steel mill", https://bigriversteel.com/wp-content/uploads/2017/12/MPT-Int.-2017-6_BRS_SMS-group_Learning-steel-mill.pdf
105. "SSAB Americas brings aboard Noodle.ai", October 23, 2019, https://www.aist.org/news/steel-news/2019/october/21-25-october-2019/ssab-americas-brings-aboard-noodle-ai/
106. "Feature: Nucor CEO Ferriola says company aspires to be the 'Amazon' of steel", https://www.spglobal.com/commodityinsights/ko/market-insights/latest-news/metals/101719-feature-nucor-ceo-ferriola-says-company-aspires-to-be-the-amazon-of-steel
107. "Financial information for the second quarter of the fiscal year ending March 31, 2022", *NS Solutions, Nippon Steel*, https://www.nssol.nipponsteel.com/en/ir/pdf/FY_Mar_2022_2Q_Financial_Information.pdf

108. Yoshitaka Kobayashi and Senoo Mitsutoshi, "AI/Big data analysis platform in JFE engineering", *JFE Technical Report*, No. 26, March 2021, https://www.jfe-steel.co.jp/en/research/report/026/pdf/026-12.pdf
109. "Made in China 2025 factsheet", https://www.plattform-i40.de/IP/Redaktion/EN/Downloads/Publikation/China/MIC2025_factsheet.pdf?__blob=publicationFile&v=6
110. D. Zhou, K. Xu, Z. Lv, J. Yang, M. Li, F. He, and G. Xu, "Intelligent manufacturing technology in the steel industry of China: A review", *Sensors*, Vol. 22, 2022, p. 8194, https://doi.org/10.3390/s22218194
111. "Primetals technologies to supply industry 4.0 package for Baosteel hot strip mill", London, June 22, 2017, https://www.primetals.com/press-media/news/primetals-technologies-to-supply-industry-40-package-for-baosteel-hot-strip-mill
112. "Baosteel—Angang high level exchange, co-consulting on steel industry transformation by internet thinking", June 12, 2015, https://www.baosteel.com/group_en/contents/2863/79090.html
113. "Sizing the prize, what's the real value of AI for your business and how can you capitalise?", https://www.pwc.com/gx/en/issues/analytics/assets/pwc-ai-analysis-sizing-the-prize-report.pdf

Index